Making Connections

Time Warner Cable and the Broadband Revolution

Making Connections

Time Warner Cable and the Broadband Revolution

A list of photo credits may be found on p. 204.

First Edition
Printed in 2011

ISBN
978-1-882771-29-5

Library of Congress Control Number
2011932285

Published by
Time Warner Cable Inc.
New York, NY
www.timewarnercable.com

Produced by
The History Factory
14140 Parke Long Court
Chantilly, Virginia 20151
www.historyfactory.com

Designed by
Pivot Design, Inc.
www.pivotdesign.com

This book was commissioned by Time Warner Cable and authored by The History Factory. The authors drew on, among other sources, interviews of past and present employees, other individuals, and publicly available sources. As such, the book does not represent the official views of Time Warner Cable; rather, it is a narrative developed by the authors from these various sources. Furthermore, the views of individuals quoted in this book are their own, and do not necessarily reflect the views of the company.

The book does not purport to be a complete or accurate account of every event in the company's history. The authors have selected which events to describe and have abbreviated certain details in describing those events. Often, the descriptions in the book rely on the recollections of individuals of events that transpired years ago. For additional information on publicly disclosed events, readers should rely on public disclosures that have previously been made.

Caution Concerning Forward-Looking Statements
This document includes certain "forward-looking statements" within the meaning of the Private Securities Litigation Reform Act of 1995. These statements are based on management's current expectations or beliefs and are subject to uncertainty and changes in circumstances. Actual results may vary materially from those expressed or implied by the statements herein due to changes in economic, business, competitive, technological, strategic and/or regulatory factors, and other factors affecting the operations of Time Warner Cable Inc. More detailed information about these factors may be found in filings by Time Warner Cable Inc. with the SEC, including its most recent Annual Report on Form 10-K and any subsequent Quarterly Reports on Form 10-Q. Time Warner Cable Inc. is under no obligation to, and expressly disclaims any such obligation to, update or alter its forward-looking statements, whether as a result of new information, future events, or otherwise.

Contents

Welcome to the History of Time Warner Cable

When I was asked to become CEO of Time Warner Cable in 2001, I immediately focused on how to cope with ever-growing competition, the organizational turmoil surrounding the merger of AOL and our parent, Time Warner, and a perception among investors that the cable industry's best days were behind it. The last thing on my mind was to sponsor the creation of a history of our company.

Somewhere along the line I realized that some of my fellow employees who had been working for Time Warner Cable for "only" five or 10 years wondered how we got where we were and why certain decisions had been made in the past. It was those questions that convinced me it would be a good idea to create a history of our company and, by necessity, a reference to the history of the entire cable industry. Some people will find this history to be of interest, and I'm sure others will wonder why in the world anyone bothered to write it down. My hope is that this small volume and accompanying online materials are able to satisfy the former without offending the latter.

I want to thank all of the people who participated in creating this history. People who were involved many years ago and people who are still involved today all took the time to share the stories and experiences that are included.

Despite the skeptics, Time Warner Cable and the cable industry have thrived during the last 10 years. The ability to do well in the face of opposition and disbelief has been a part of the cable industry ever since it started over 60 years ago. We have done that by always understanding consumer trends and changing technology and working to put the two together to develop products and services that people want. Of course, the business has changed, and it will continue to change as the world around us does. It has gotten harder to change as we have gotten more mature and larger, but we know that creating things people want is both fun and profitable. With that spirit I expect that Time Warner Cable and the industry it is a part of will be around for many, many more years. It will change in ways that are unpredictable, and perhaps it will be unrecognizable 10 years from now. However, whatever form it takes, the direct line of entrepreneurial enthusiasm depicted in this history will continue!

Sincerely,

Glenn Britt
Chairman and Chief Executive Officer

Experience it. Use your smartphone to scan this QR code and visit Time Warner Cable's heritage website, history.timewarnercable.com

Time Warner Cable Milestones

American Television and Communications (and predecessor companies)

Warner Cable (and predecessor companies)

Time Warner Cable

Government and Industry

Late 1940s First community antenna television (CATV) systems built in Oregon and Pennsylvania

1948-1952 Federal Communications Commission (FCC) freeze on new broadcast TV licenses fuels CATV growth

1950 Franklin, Pennsylvania, system built by Jack and Abe Harter

1951 National Community Television Association (NCTA) formed

- Pottsville, Pennsylvania, system built by Martin Malarkey

1953 Bill Daniels enters cable business, building his first CATV system in Casper, Wyoming

1960 TelePrompTer enters cable business

1962 Television Communications (TVC) formed

1963 Monty Rifkin, formerly with TelePrompTer, joins Bill Daniels in Denver to own, operate cable systems

- Alan Gerry forms Cablevision Industries (CVI)

1966 Cypress Communications formed

Late 1960s Time Inc. buys stake in Manhattan-based Sterling Communications and other cable systems mostly in Midwest and West

1968 Bill Daniels and Monty Rifkin co-found American Television and Communications Corp. (ATC) on June 4—planting the roots of today's Time Warner Cable

- Monty Rifkin named ATC chairman and CEO
- FCC bans ownership of cable and broadcast systems in same market

1969 ATC goes public

1970 ATC acquires systems including Albany, New York, and wins the Orlando, Florida, franchise

1972 Proposed merger of ATC and Cox announced (rejected by Justice Dept. in 1973)

- ATC buys Jefferson-Carolina Corp.
- HBO debuts
- Warner Communications buys TVC, cable assets of Continental Telephone, and Cypress Communications
- Completes two-way Akron, Ohio, system

1973 Time Inc. acquires 9 percent of ATC

- Warner Cable subsidiary formed (140 systems in 31 states, 453,000 subscribers)

- Gus Hauser named Warner Cable CEO
- Time Inc. acquires all of Sterling Manhattan (renamed Manhattan Cable)

1975 HBO carries "Thrilla in Manila" heavyweight fight via satellite, sparking growth of HBO and cable

1976 Ted Turner's superstation (later renamed WTBS) launches on satellite

- Columbus, Ohio, system built

1977 QUBE, the first two-way interactive cable system, launches

1978 Time Inc. agrees to buy 100 percent of ATC

1979 Warner and American Express form Warner Amex cable partnership, with 700,000 subscribers

- ATC hits 1,000,000 subscribers
- Brian Lamb launches C-SPAN
- Warner Communications launches Nickelodeon

1980 Turner launches CNN

- Westinghouse cable unit Group W buys TelePrompTer

1981 Warner Communications launches MTV

1982 Trygve Myhren named ATC chairman and CEO

1983

Drew Lewis named Warner Amex chairman and CEO, moves company headquarters from New York City to Blue Bell, Pennsylvania

1984 Charlotte and Greensboro, North Carolina, systems acquired

- Warner Amex sells systems in Pittsburgh, suburban Chicago, suburban St. Louis to cut costs (1.2 million subscribers following deals)
- Cable Communications Policy Act of 1984

1985 ATC makes Group W, Paragon deals (adds systems in Arizona, California, Florida, Maine, New Hampshire, New York [including northern Manhattan] and Texas)

- Warner buys out Amex partnership interest, sells cable networks including MTV and Nickelodeon to Viacom
- Warner Cable company headquarters moves to Columbus, Ohio

1986 ATC publicly lists stock for second time

- Jim Gray named Warner Cable president

1987 -1988 ATC develops fiber-optic cable network

1987 Warner launches 100+ channel cable system in multiple test areas

1988 Joe Collins named ATC chairman and CEO

- ATC company headquarters moves from Denver to Stamford, Connecticut

1989 Time-Warner merger announced, bringing together the companies' 6.1 million cable subscribers (although ATC and Warner Cable would continue to operate separately and retain their names until 1992, when ATC is no longer a public company)

- Joe Collins named Time Warner Cable CEO

1991 PrimeStar satellite service jointly launched by Time Warner Cable, other cable companies

1992 ATC and Warner Cable combined into Time Warner Cable with a total of 7.1 million subscribers

- Time Warner Entertainment (TWE) partnership formed, includes Time Warner Cable
- NY1 News launches in New York City
- Cable Television Consumer Protection and Competition Act of 1992

1993 Circuit-switched telephony tests in Rochester, New York

1994 Emmy® Award for work in broadband fiber-optic technology

- The Full Service Network (FSN), a digital system that is an early forerunner of video on demand and other innovative services to come, is tested in Orlando

1995 Test in Elmira, New York, of Line Runner high-speed Internet service (later rechristened Road Runner®)

- Time Warner Entertainment-Advance/Newhouse (TWE-A/N) partnership formed

1996 Road Runner®, the first cable-delivered high-speed Internet service, starts national rollout in Akron and Canton, Ohio

- CVI, KBLCOM, Summit Communications systems acquired in '95 and '96
- 1996 Telecommunications Act

1998 Road Runner® Business Class (later named Time Warner Cable Business Class) begins offering video and high-speed data services to small businesses

1999 Digital cable rolls out company-wide

- Video on demand launches
- System exchanges with AT&T subsidiaries, MediaOne, Fanch, and Comcast

2000 AOL-Time Warner deal announced

2001 Glenn Britt named Time Warner Cable CEO

2002 Restructuring of TWE-A/N

- High-definition TV launches
- Integrated DVR launches

2003 Launch of digital phone service via Voice over Internet Protocol (VoIP) technology

2005 Start Over® digital video debuts (later wins an Emmy® Award)

- Rollout of triple play offering of video, high-speed data, and phone

2006 Adelphia-Comcast transactions add 3.3 million basic video subscribers in areas including Los Angeles and Dallas (more than 14.5 million subscribers total)

2007 Time Warner Cable stock publicly traded on New York Stock Exchange

- Look Back™ debuts
- Time Warner Cable Business Class launches Business Class Phone
- Headquarters moves to New York City

2008 Emmy® Award for video on demand contributions

2009 Separation of Time Warner Cable from Time Warner Inc.

2010 TV Everywhere online video launches (with ESPN)

2011 TWCable TV™ app for the Apple iPad® launches

- Time Warner Cable acquires business services subsidiary NaviSite to enable "cloud" services
- Time Warner Cable agrees to pay $3 billion for Midwestern cable operator Insight Communications Company Inc., which has 750,000 subscribers

With broadcast television reception hard to come by in the shadow of Mt. Hood, the residents of The Dalles, Oregon, welcomed the arrival of coaxial cable installers, and cable television, in 1954.

Chapter 1

Birth of an Industry

Late 1940s–Late 1960s

TelePrompTer, one of the first large companies to invest in cable television, made its name as the inventor of machines that provided television performers, such as these soap-opera actresses, with easy-to-read lines that scrolled down as their scenes progressed.

Monroe "Monty" Rifkin had a choice to make one afternoon in the spring of 1959. He could finish the sandwich on his desk in the TelePrompTer Inc. offices near New York's Times Square. Or he could meet the promoter from Denver waiting in the outer office to speak with whomever was in charge about something called community antenna television.

Rifkin had no idea what was at stake as his appetite for lunch vied with his appetite for new business. His choice set in motion a cross-country, multi-generational personal and business odyssey—replete with larger-than-life characters, jobs created by the tens of thousands, corporate crises, and culture-changing technologies. A key turning point in that odyssey was Rifkin's lead role nearly a decade later in creating and taking public American Television and Communications Corporation, known as ATC, to which Time Warner Cable traces its operating company roots.

The decision was not as clear-cut as hindsight might suggest. First, the Gaiety Delicatessen on 47th Street made an excellent pastrami on rye. Second, Rifkin had grown tired of smooth-talking promoters—TelePrompTer CEO Irving B. Kahn in particular.

Monroe "Monty" Rifkin rarely ventured farther west than the Times Square offices of his employer, TelePrompTer, until, at the urging of industry legend Bill Daniels, he flew to Denver in 1959 to get a firsthand look at early cable television systems.

TelePrompTer CEO Irving Berlin Kahn took Monty Rifkin's advice and bet his company on Community Antenna Television (CATV), later becoming an influential, as well as infamous, industry figure.

Rifkin was the man in charge at TelePrompTer that afternoon because Kahn (who always found an excuse to remind those in earshot that he was named after his uncle Irving Berlin) was having one of his elaborate, extended lunches, probably with some of the colorful characters associated with the fights TelePrompTer had been showing in movie theaters across the country via closed-circuit television. Between drinks, Kahn spun tales of TelePrompTer's potential beyond the "idiot boxes" the company created, which scrolled lines of dialog for television actors and announcers to read on-air.

An accountant by training and temperament, Rifkin prayed that Kahn's stories bore at least a semblance of reality to the cash-strapped enterprise Rifkin, still several months shy of his 30th birthday, was effectively running as executive vice president. Kahn may have been a visionary, but he had a habit of not letting facts get in the way of a good story.[1]

"What's That?"

Despite his reservations, Rifkin recalled liking Bill Daniels almost from the moment the stocky Westerner, with his signature tailored suit and pocket handkerchief, strode across the office, shook his hand, and got right to the point. He had been referred to Kahn and TelePrompTer by *Television Digest* editor Marty Kordell.[2] Daniels had read about the company's closed-circuit televised fights and wanted to talk to TelePrompTer about a related opportunity: community antenna television, or CATV. "What's that?" asked Rifkin, who like most Americans, especially those living in cities and towns served by broadcast television, had never heard of CATV.

Daniels explained that roughly a half-million homes in the United States were receiving television through systems connecting giant antennas, usually built on hilltops, to homes via what was known as coaxial cable. These customers were either too far from broadcast television station antennas to receive adequate signals, or the signals were blocked by mountains or hills. Entrepreneurs—many of them appliance and television salesmen or radio engineers—had built hundreds of small CATV systems across the country. The first systems, built in Pennsylvania and Oregon, were barely a decade old, if that. Daniels himself was a joint owner of several systems and had recently set up shop as Daniels and Associates.

The technology pitch was intriguing, but what especially interested Rifkin was the economics of the fledgling industry. Not surprisingly, Daniels argued that the CATV industry presented a great buying opportunity. Daniels also didn't spare the flattery, pointing out that an investment from a sophisticated, publicly traded company like TelePrompTer would raise the industry's mom-and-pop profile, make banks and insurers more willing to lend to system owners, and in turn increase the value of CATV systems.

Rifkin asked for more financial information, which Daniels supplied during two subsequent visits to New York. On the third visit, Daniels pressed Rifkin for a decision. "Monty, I can talk to you here in New York all week long, but you've got to come out and see some of these companies."[3] Rifkin, who had never been farther west than Chicago, finally agreed, booking a flight that landed him at Denver's Stapleton Airport in August 1959.

Experience it. Use your smartphone to watch a video of Monty Rifkin describing what it was like to meet charismatic cable legend Bill Daniels.

TOP

First-generation cable technicians pose proudly before the looming, if rudimentary, antennas that received television signals from broadcast stations in Pennsylvania and Ohio and carried them to customers of the Meadville, Pennsylvania, system.

LEFT

Bill Daniels was the first cable system operator to fully realize the business potential of the fledgling industry and was instrumental in attracting financing from outside sources, including TelePrompTer's leadership team of Irving Kahn and Monty Rifkin.

TELEVISION
RECEIVERS
ARE DISPLAYED THROUGH
THE COURTESY OF THE
FOLLOWING Television DEALERS
MEMBERS OF THE
MEADVILLE RETAIL COUNCIL
MONTGOMERY-WARD
BOB'S HOME RADIO
L.G. WINELAND'S APPLIANCE
GOODYEAR SERVICE STORE
FIRESTONE STORE
W.T. GRANT STORE
WESTERN AUTO STORE
AMPLIFIER
AMPLIFIER HOUSING
TELEVISION

1959

Meadville, Pennsylvania's Master Antenna Service brought television to residents of the towns lining the river valleys of the northwestern part of the state, where broadcast signals failed to reach.

TOP
Feisty Philo T. Farnsworth battled with his scientific rival Vladimir Zworykin for the right to the title of "father" of television.

RIGHT
Then–U.S. Commerce Secretary Herbert Hoover, in Washington, D.C., was featured in a 1927 television demonstration originating from AT&T's Bell Laboratories in New York. The New York Times *doubted the technology's commercial use.*

Broadcast vs. Wired

Wired television systems were a novelty item in the postwar period, which was dominated by the broadcast networks. But that fact doesn't convey a fully accurate picture of the development of this truly revolutionary medium. As early as the late 1800s, television as a technological concept was in wide circulation among scientists and engineers in Europe and the United States. The word *television* itself appears to have been first used in 1900. And AT&T was studying the transmission of visual images over wired systems as early as the 1890s. In fact, it was years before wireless transmission was considered remotely feasible.[4]

By the 1920s, television technology had made significant strides—despite the long-running patent disputes between the rival "fathers" of television, feisty independent Philo Farnsworth and RCA-backed Vladimir Zworykin. AT&T gave a public demonstration of television in 1927 at Bell Laboratories in New York, with then–U.S. Commerce Secretary Herbert Hoover speaking in Washington. A front-page article in *The New York Times* described the technology as "a photo come to life," while adding, "commercial use in doubt." The rudimentary television signal was carried on phone lines and via a radio link.[5]

As improved television images required more bandwidth, AT&T turned its attention by the late 1920s and '30s to coaxial cable, an early version of which had been used for high-volume trans-Atlantic telegraph lines. As opposed to the "twisted pair" of copper wires comprising the phone company's voice circuits, coaxial cable consists of a single copper wire running down the middle of a "pipe" composed of copper or another metal. Holding the central wire in place in the tube is an insulating material such as hard rubber or polyethylene spacers or discs. This coaxial configuration—the central wire axis within the axis of the outer tube—provides for transmission of much greater bandwidth than traditional twisted pair circuits, especially at the higher frequencies that characterize television signals. A Federal Communications Commission 1938 staff report on the possible development of television in the United States highlighted the role coaxial cable could play. The U.S. television system could "develop into some sort of wire plant transmission utilizing the present basic distribution network of the Bell System, with the addition of coaxial cable or carrier techniques now available or likely to be developed out of the Bell System's present research on new methods of broad-band wire transmission."[6]

RCA and its commanding leader, David Sarnoff, had another idea. Sarnoff had spent years acquiring television patent rights and developing television broadcast standards. The overwhelming success of free-to-the-consumer, advertising-supported radio broadcasting and its system of networks, as well as RCA's commanding role in the development of that technology, made it obvious, and imperative, to the Sarnoff camp that television follow suit. Coaxial cable would be used as the backbone to carry television signals from network studios to affiliates.[7]

Sarnoff declared the inauguration of broadcast television at the 1939 World's Fair in New York. Roughly 30 stations were launched by the summer of 1941. It was a short-lived boom. The Japanese attack on Pearl Harbor that December brought an almost immediate halt to the construction of new stations, or non-military production of related equipment, as America entered WWII. Engineers working in this emerging field were recruited by the hundreds for radar and other military technologies. Only six television stations remained on the air through the war years in the States.[8]

If the rival means of transmitting television signals in the early developmental years of the medium can be seen in one sense as pitting Ma Bell's coaxial cable against RCA's broadcast standards, RCA clearly emerged victorious. It, or at least broadcast technology, remained so for decades. The FCC staffers' vision of a wired broadband nation, however, wouldn't die. It was nurtured—slowly and in fits and starts around the country, sometimes with the aid of regulators, sometimes in the face of regulatory constraints—in the form of community antenna, or as it would later be renamed, cable television.

RCA leader David Sarnoff declared the dawn of commercial television in the United States at the 1939 World's Fair in New York. America's entry into WWII in December 1941 effectively halted the spread of the long-awaited technology beyond a handful of broadcast stations until the war's end.

TOP
Milton Berle and other radio personalities made the transition to television and helped spark tremendous public demand for the medium in the immediate postwar years. A regulatory "freeze" on new broadcast television licenses lasting from 1948 to 1952 created an opportunity for pioneering cable operators to gain a toehold in the business.

RIGHT
Federal Communications Commission members, shown here visiting a CATV antenna site, were uncertain how, or even whether, to regulate cable television in its early years.

"The Freeze"

With war's end, it was finally time for television, the medium of the future for the previous two decades, to make its sustained commercial debut. In late 1947, 17 stations were on the air. A year later, that number had jumped to nearly 50. *Radio and Television News* declared 1948 the "Year of Television."[9]

The popular press was all but obsessed with the new technology, even if viewers comprised less than one-half of 1 percent of households and remained clustered around major urban markets, led by New York, Los Angeles, and Chicago.[10] The coverage also conveyed a sense of uncertainty about the impact of the new technology on American society and, at least by implication, on existing communications media, including newspapers, magazines, radio, and film. *Newsweek* described television in October 1948 as a "locomotive on the loose."[11] *Time* weighed in the following January with an article on television titled, "Young Monster."[12]

The growing number of television broadcast stations was impressive yet manageable, as far as the FCC was concerned. It was what was coming down the pipeline that had regulators worried. From 30 applications for television broadcast licenses in October 1947, the demand soared tenfold to 300 by October 1948. Profits remained elusive. Not even RCA's National Broadcasting Company subsidiary was making money in television broadcasting at that point.[13]

More important, earlier FCC decisions allocating television spectrum to the 12-channel very high frequency (VHF) band was leading to signal interference, as experts had predicted, from channels being broadcast from stations in different cities—in some cases, as far apart as

"Abel Cable" was an industry mascot created by the National Community Television Association (NCTA) to convey cable television's friendly, can-do ability to deliver television signals to communities beyond the reach of broadcast television.

Detroit and Cleveland. How to apply ultra-high frequency (UHF) spectrum to television puzzled consumer regulators and technical experts for years.[14]

To address these growing pains, the FCC declared a freeze on the granting of additional broadcast licenses, effective September 30, 1948. Expected to last six months or so, the freeze was in effect for four years as agency officials worked through the issues besetting television broadcasting. The only stations allowed to operate were those 108 that had been granted operating licenses prior to the restrictions. And nearly all of those were in or near major cities.

Regulatory Umbrella

What became known as "The Freeze" might in another sense just as accurately have been termed the Pro Community Antenna Television Act. The halt on launching new stations had the effect of whetting the public's appetite for television, not dulling it. If this new medium was being rationed by the government—much like sugar and nylons had been during the war—it must be good. In fact, by the time the freeze was lifted in 1952, television's household penetration had gone from less than one-half of 1 percent to a stunning one in three U.S. households.[15]

A portion of that increase can be attributed to mounting economic prosperity as the nation entered the boom years of the 1950s.

By the late 1950s, cable technology enabled hundreds of thousands of Americans, including this woman in Bartlesville, Oklahoma, to enjoy the new television medium that was rapidly redefining postwar American culture.

Cable antennas, initially referred to as community antennas, took different forms depending on the manufacturer and demands of the local systems. This antenna, pictured in 1955, was located in Holmdel, New Jersey.

Demand for television was soaring, despite broadcast stations being limited to major urban markets during this period. That had the effect of cementing NBC and rival Columbia Broadcasting System as the leading broadcast networks, and forced latecomer American Broadcasting Company to play catch-up for years. A fourth network, Dumont, never recovered from the effects of the freeze and faded to black a few years later.

Enter community antenna television. The freeze effectively barred the networks and independent broadcasters alike from venturing beyond the top 20 to 30 U.S. urban markets, where most stations were clustered and applications for new stations focused. But since community antenna systems simply received a signal and did not broadcast at the time, they weren't regulated by the FCC and therefore weren't constrained by the freeze. As much as the industry and regulators locked horns over CATV oversight in coming years, there is no question that the freeze on broadcasters provided community antenna entrepreneurs with a protective regulatory umbrella under which they could build their systems.

Who's on First?

Who invented community antenna television? After decades of claims and counter-claims, there remains no clear-cut answer to the question. In addition to a dearth of documentation supporting claims of some of the early practitioners, there is also a definitional problem: What, exactly, constituted a community antenna system?[16]

With all the discussion of who was the "father" of cable television, the more pertinent question might be, who was the mother? Grace Parsons may be more deserving of the title than anyone. The Astoria, Oregon, native was visiting Seattle in the fall of 1948 when she saw her first demonstration of television, conducted by KRSC-TV. She liked what she saw. She returned home and told her husband, Leonard, who had been building radios for years and owned local radio station KAST, that radio was no longer enough for her. "I want pictures with my radio," she declared.[17]

Given the fact that Seattle was 125 miles away, Parsons told his wife it was technically impossible. Besides, his ultimate goal, shared by independent radio station owners around the country, was to launch a broadcast television station as soon as the FCC freeze was lifted. She persisted; he began experimenting.

Parsons was personal friends with the television station manager in Seattle, Robert Priebe, who let him know when the Seattle station would be testing its signal. Using a homemade testing device, Parsons eventually located the signal from a perch on the rooftop of the Astoria Hotel, erected an antenna, and ran a wire to his nearby apartment building. The Parsonses and friends were watching television on a set with a nine-inch screen in Astoria on Thanksgiving Day 1948, when KRSC officially went live with its broadcasts. By the summer of 1949, Parsons had roughly 30 homes and businesses on the system, with people placing amplifiers to boost the signal in their attics or apartments. Coverage of the system by the Associated Press in the summer of 1949, and an article in *Popular Mechanics* in April 1950, helped spread the word about community antenna television across the country. Parsons, unclear as to what authority if any held regulatory sway over such a system, also alerted the FCC to his system.[18]

Ed Parsons' Astoria, Oregon, community antenna system received early recognition, and a memorial, for its pioneering role in the development of cable television.

Did coaxial cable running from a hilltop antenna to television sets in a local barroom, pool hall, or appliance store or two constitute a system? Whether it did or not, one of the first and more amusing instances of community outreach on the part of a CATV operator—which became critical to the industry's success in coming years—is believed to have occurred in 1947 or 1948 in Mahanoy City, Pennsylvania. Ed "Peanuts" Trusky erected a tower on nearby Broad Mountain and brought television signals from Philadelphia into his pool hall, and thereby the town. Perhaps aware of the negative associations often heaped on such establishments, Trusky reached out to the local church three doors down. He extended the antenna cable to the rectory for the enjoyment, and of course edification, of local clergy.[19]

Trusky was typical of the numerous individual bar and appliance store owners who, in the late 1940s and early '50s, were testing community antenna setups in the hills of eastern Pennsylvania. About 100 miles or so from Philadelphia and its three main broadcast stations, which had been licensed before the war, the coal mining region's hilltop towns could easily receive signals from Philadelphia. It was the many towns in the valleys that created demand for community antenna television.

John Walsonovich (who later shortened his last name to Walson) was believed to have been a patron of Trusky's pool hall. He may have developed a taste for the new medium there, and launched a commercial community antenna company in Mahanoy City in late 1950 or early 1951.

The valley towns in the rugged western portion of Pennsylvania—especially along the Allegheny River, whose hilltop neighbors could receive television signals from Pittsburgh and Erie—were likewise creating demand for community antenna television. Jack and Abe Harter, who ran the Harter Brothers appliance store in Franklin, built a hilltop antenna and began running wire down into the valley toward town in the summer of 1949 or early 1950. By that summer, they had a five-mile line with 40 customers paying $9.25 a month.[20] A few years later, they restructured the business and joined forces with other cable entrepreneurs in town.

Back in the eastern portion of the state, Robert Tarlton, who had visited the Harters in the summer of 1950, had returned to Lansford, about 90 miles from Philadelphia. With backing from local businessmen and friends, including George Bright, he formed Panther Valley Television Company. The appliance-store owner, who had run wire down from a hilltop antenna to demonstrate television set performance in his store as early as 1947, began running wires down from a new 85-foot tower in late 1950 to build a system in Lansford.[21]

Tarlton's company was ordering so many amplifiers from Jerrold Electronics in Philadelphia that owner Milton Jerrold Shapp, who later served as governor of Pennsylvania from 1970 through 1978, sent a salesman up to Lansford to see what was going on. Shapp himself later paid a visit and realized that he could make a business out of designing and selling equipment geared specifically for community antenna television use.

Martin Malarkey founded the cable system in Pottsville, Pennsylvania, and also led the creation of the National Community Television Association (NCTA), serving as the industry group's first leader. "Cable" was later substituted for "community."

Shapp subsequently used the Lansford system as a testing ground for new equipment design. Articles about the close working relationship between the television system and the equipment maker appeared in *The New York Times*, *Wall Street Journal*, *Time*, and *Newsweek* in late 1950 and early 1951. The publicity helped trigger a burst of construction of community antenna systems and raised awareness of the new industry.[22]

Martin Malarkey, who managed his family's Pottsville, Pennsylvania, store selling musical instruments, radios, and the occasional television set, got the idea for a community antenna system after watching television in a room at the Waldorf-Astoria Hotel in Manhattan in 1949. Building on a model widely used for radio, the hotel had constructed an antenna on its roof and delivered television to individual guest rooms. A handful of apartment building complexes in New York and Chicago had taken similar master antenna approaches to distributing television to tenants by this time as well.

Malarkey met with the hotel's chief engineer and later engineers at RCA who had designed the system. He had his own system up and running using RCA equipment by 1951 and worked as a consultant to other system owners, including Bill Daniels. He also served as the first president of the industry trade group, the National Community Television Association (NCTA), which he and a handful of other early cable industry builders formed that same year in Pottsville's Necho Allen Hotel.[23]

The NCTA held its first meeting in 1951 in the Necho Allen Hotel in Pottsville, Pennsylvania.

Cable TV Spreads From Rural Regions to Cities

Cable television systems, once regarded as a small town phenomenon, exist today in the heart of New York City and in areas of Los Angeles, San Francisco, San Diego, Cleveland and Austin, Texas. Franchises have been granted or applied for in Philadelphia, Pittsburgh, suburban sections of Boston and Chicago — in fact, most of the nation's major cities.

The subscriber potential of a handful of such markets could, virtually overnight, jumble all the growth patterns of an industry that has until recently served the modest TV reception needs of places like Lansford, Pa.; Cut Bank, Montana; Junction City, Kansas; and Painted Post, New York.

What is it that brings CATV—or community or cable TV—from the rural hills to the urban city with such force?

CATV originally helped viewers in outlying areas receive TV signals otherwise not available to them, either because of the distance from the broadcast stations or interference from surrounding mountains.

But the cities have their problems too. Many New York City viewers, for example, have very poor reception although broadcast stations are within walking distance. Steel and glass skyscrapers act just like mountain ranges to hinder good reception.

Sensitive CATV antennas can receive the broadcast signals and transmit them clearly and sharply to a subscriber's apartment or home via cable. Color television, with its technically exacting reception requirements has greatly increased big city interest in cable TV facilities.

A typical CATV system, with 12 channels to be filled, can offer a rich choice of viewing fare — from entertainment and educational programs to news and weather channels and local interest sports or politics. Special programming for specific neighborhood or ethnic groups is another possibility.

Systems already have been devised with as many as 20 channels and a virtually unlimited number are technically possible. This "super-highway" of information, it is predicted, may someday bring direct hookup from homes to computers to do income taxes and library research.

Also possible are electronic newspapers, voting and opinion surveys and gas meter reading through the predicted cable-computer system.

CATV enters the big city because it is needed. And the big city offers CATV an unparalleled opportunity to grow and develop as a part of the total communications system of the future.

By the late 1960s, cable television, often offering a number of independent channels as well as the three major broadcast networks, was expanding beyond its rural roots and beginning to make headway in penetrating smaller cities.

"A Hell of a Potential Business"

Bill Daniels is widely credited with creating a legitimate industry out of do-it-yourselfers stringing coaxial cables down back alleys. "I was the first guy to recognize it as a hell of a potential business," Daniels would later claim. "And I brought the financial community in to really make it a business rather than an extension of an appliance shop."[24] Luring Rifkin out West in order to get TelePrompTer to invest in the industry was an important early step in that process.

A decorated fighter pilot and veteran of both WWII and the Korean War—receiving the Naval Cross, Air Medal, and Distinguished Flying Cross, as well as flying on the demonstration team that was the precursor of the famed Blue Angels—Daniels was looking for new challenges once he returned to civilian life in 1952. He was slightly bored by his job as an insurance broker to the oil industry in New Mexico. He had also had a falling-out with his brother, who had run the agency started by their father while Daniels was in the service, and wasn't too keen on letting his brother back into the business on an equal footing. Daniels stopped in at a Denver bar one day en route to Casper, Wyoming, where he hoped to find new insurance clients among the state's thriving oil business. He witnessed television for the first time at age 32, and it was a revelation. "My God, what an invention this is," he thought to himself.[25]

Gene Schneider was one of dozens of cable industry pioneers who credited industry leader Bill Daniels with helping them get a start in the cable television business.

Daniels built an insurance business in Casper to pay the bills and support his family. But he couldn't get television off his mind. The allure was especially strong since Casper residents didn't have a broadcast station of their own, and the locale was too far from Denver, the nearest city of any size, to pick up signals.

By 1953, after learning of community antenna television, Daniels was determined to bring the new medium to Casper. Bouncing the idea off some friends in the oil business, he was put in touch with Richard and Gene Schneider, two Army-trained engineers. They agreed to be the system operators, as soon as Daniels managed to put together a system. Jerrold Electronics' marketing chief, Zal Garfield, learned of Daniels' interest, discussed the idea with him, and directed him in turn to Martin Malarkey for tips on setting up a community antenna business.

Malarkey didn't have any qualms about sharing his trade secrets. That is, as long as the person seeking information was willing to pay his consulting fee of $500 a day. Daniels didn't blink at the price and jumped on a plane to Pennsylvania for the opportunity to pick Malarkey's brain for two days.[26]

Daniels returned to Casper and, with just $5,000 of his own money, raised an initial $125,000 from a group of local oilmen and a prominent lawyer. Borrowing from their local banks, they raised another $250,000 in working capital. The business plan was typical of early CATV systems and was based on a one-page permit, not a franchise agreement, from the Town of Casper granting Daniels permission to run cable down streets and alleys.

"We were charging 150 bucks per connection because we had a monopoly," Daniels said. "With every 150 bucks we got, we would build a couple more blocks of plant down the alley to get more customers. And we charged $7.50 a month for one channel." He christened the enterprise Community Television System of Wyoming.[27]

His system was patterned after Malarkey's business model, but Daniels could claim a CATV "first" in terms of signal transmission. Even with no intervening mountains, a television signal can be transmitted only so far on a "line of sight" basis due to the curvature of the earth. Having to cover close to 200 miles to reach Casper with the signal, Daniels, who had worked with radar systems in the service but otherwise had no formal engineering training, hit on the idea of using microwave transmitters and relay towers to carry the signal. Working with Jerrold, he leased a microwave system from the Bell System, which included a relay tower in Laramie, Wyoming, and the system worked flawlessly. Within a few years, microwave transmission of television signals was in use at a number of CATV systems.

Daniels proceeded to build systems in Rawlins, Wyoming, and Farmington, New Mexico, once the Casper system was operational. He retained equity stakes in each system but was barely making ends meet. Plowing all of his money back into the next system, Daniels also needed to keep his insurance business going to pay the bills. Yet he was increasingly convinced that he could make a living in the cable business.

NCTA

Daniels may have been operating in towns out West that many of his cable compatriots back in Pennsylvania would be hard-pressed to find on a map, but he was very much on their collective radar. His contacts with Jerrold and continuing equipment orders, and his discussions with Malarkey, made him a man to watch among cable system operators by the mid-1950s. He was so successful, starting to build three systems in as many years, that in 1956 he was tapped to succeed Malarkey as the second president of the National Community Television Association. "Cable" later substituted for "Community."

Then, as now, the trade association served as a forum for sharing common interests and collectively addressing public policy challenges to the industry's growth. Many of the key regulatory and competitive issues facing the association and the industry in its infancy remain central to the NCTA's efforts today. Of course, the specifics have evolved with the industry and television technology.

A challenge from the Internal Revenue Service threatened to derail industry growth from its earliest days, and actually led to the formation of the NCTA. In 1951, an IRS agent knocked on Bob Tarlton's door at Panther Valley Television Company and informed him that his community antenna system was subject to an 8 percent federal excise tax as a "wire and equipment service." Similar taxes were placed on stock quotation systems and telephone, telegraph, cable, and radio service providers. But with so many systems operators barely getting by month to month, using subscriber fees to pay for cable and other equipment purchases, it was feared that several might be forced out of business.[28]

Tarlton contacted Malarkey, and he in turn pulled together the initial group that formed the NCTA. Representatives from eight systems in Pennsylvania were included.

The initial approach to fighting the tax issue was to focus on upfront installation fees, which played a pivotal role in financing further system growth. The industry argued that they should be treated as "contributions in aid of construction," which would exempt them from income taxes, as opposed to ordinary revenue. There was a precedent in rural utility accounting from the 1920s.

Early Legal Strategy

In the process of addressing the broader issue of whether all industry revenues would face the 8 percent excise tax, the trade association concluded by 1953 that it needed legal representation in Washington. At the same time, E. Stratton Smith, an attorney in private practice who earlier had been on the FCC staff when it had received notice from Leonard Parsons in 1949 about his Astoria, Oregon, community antenna system, concluded that he would like to represent the NCTA. With more knowledge of community antenna television than virtually any other attorney in the country, Smith had little trouble landing the association as a new client.

He was an excellent choice. Smith quickly developed a strategy to address the broad tax issue, which formed the basis for the industry's response to regulation or oversight of any stripe well into the 1960s. His argument, adopted by the industry, was that community antenna television was a completely passive system and not a service in any form. They simply provided an antenna and connecting cable to subscribers; they weren't "communicating" or generating anything.[29]

In November 1956, the U.S. Court of Appeals for the Fourth Circuit located in Richmond, Virginia, ruled in Lilly *v.* United States in the industry's favor: "We think it clear that this community antenna service was a mere adjunct of the television receiving sets with which it was connected and was in no sense a communication service or facility such as it was the purpose of the statue to tax."[30] A dozen years later, the U.S. Supreme Court cited Lilly in support of its ruling in the industry's favor in a landmark copyright dispute with television broadcasters.

The CATV industry didn't fare as well on the treatment of installation charges. The courts ultimately ruled against the industry in 1958. By that time, however, the typical system was in a much stronger financial position.

"Hell, I've Got a Management Company!"

While serving as president of the NCTA, Daniels received numerous queries from eager entrepreneurs wanting to get into the business, and some from financiers wondering if any systems were for sale. Daniels was careful not to appear to be taking advantage of his position as head of the association. But as soon as his term ended in 1958, he was convinced that he could make community antenna television into a full-time career. He moved to Denver and opened for business as Daniels and Associates, mailing mimeographed flyers to advertise himself as a consultant, broker, and investment banker to the CATV industry.[31] He was creating a new industry niche, and he dominated it for more than 25 years.

Among the first deals Daniels brokered was to find a buyer for Joe Saricks, a CATV system operator in Bradford, Pennsylvania, who wanted $1 million, cash, for his system. Daniels told him that he would get $1,050,000, with the $50,000 representing his fee. Daniels was promoting the industry at a dinner in Rapid City, South Dakota, shortly after speaking with Saricks. In the audience was Charles Sammons, a Dallas businessman and owner of Reserve Life Insurance Company, looking to diversify his holdings. After the dinner, he sent Daniels a note. "I like that business, bring me a deal."[32] Daniels quickly put buyer and seller together.

In typical Daniels fashion, he pounced on another business opportunity with Sammons, creating a new role for himself and his firm in the process. "I sold several properties to him, and his problem was he didn't know how to manage them. I said, 'Hell, I've got a management company!' Which I didn't have, but I formed it mentally that minute," Daniels recalled in a 1986 interview. He cut a deal to manage the companies and also get a 20 percent equity stake. "I earned that out and that's how I really got my equities going, because in those days, I could find people to buy properties, but there weren't guys around that could manage them. And I saw an opportunity to form a 'management company.'"[33]

TelePrompTer

Monty Rifkin represented another major potential buyer for Daniels as he touched down in Denver in August 1959. Daniels led Rifkin on a shopping trip that included visits to the CATV systems Daniels owned a piece of in Casper, Wyoming—which was co-owned and operated by Gene and Richard Schneider, who later built their own cable TV empire—and Rawlins, Wyoming, as well as Farmington, New Mexico. Another system Daniels thought might be on the market was in Silver City, New Mexico, owned by cable television pioneer Bruce Merrill, CEO of AMECO (Antenna Vision Manufacturing and Engineering Company), based in Phoenix, Arizona.

"What I found most interesting about looking at those systems was the fact that the owners were kind of hiding them in back alleys," Rifkin recalled. "They had nondescript offices, kind of unidentified, and when I questioned why, they said they really didn't want to call attention to themselves. They really felt that they were stealing. That the public was paying them for an antenna service that in other parts of the country, most parts of the country, was free."[34]

Rifkin concluded that for these and other reasons, including lack of easy access to capital to fund their growth among them, the systems were undervalued. He hadn't believed Daniels initially when he told him that a CATV system could be had for the bargain-bin price of three to three-and-a-half times annual cash flow. "You just don't buy solid businesses and get your money back that fast," Rifkin said.[35]

He had a better grasp of industry dynamics after meeting with the system owners and operators. "There was a real inferiority complex amongst those operations," he said. "They were pulling a lot of cash out of the business, were reinvesting it, were expanding their cable lines, starting to build other markets, and they were obviously running a little scared."[36]

"It's Going to be a Great Business!"

Rifkin was sold. He flew back to New York and made his pitch to Kahn: "Irving, we ought to gather every dollar we can raise. We ought to sell off any of our other businesses that we can sell, and we ought to take every bit of capital we can get or use our stock and buy these cable systems. I think it's going to be a great business!"[37]

Kahn loved the idea and was willing to go all in, while recognizing that their resources were limited. Kahn and Rifkin turned to their in-house technical expert, Hub Schlafly, before making a major bet on the industry, which would include selling their principal line of business, the TelePrompTer unit, while retaining the corporate name. As vice president of engineering, he held a patent on the creation of the TelePrompTer device.[38]

Schlafly, who had worked on military radar applications during WWII, and later for General Electric and 20th Century Fox, was concerned about the impact of satellite technology on CATV. The Soviets had launched *Sputnik* barely two years earlier, and the threat—and potential—of satellites galvanized scientists and engineers across the country and led to a renewed emphasis on science and math in the nation's schools. TelePrompTer commissioned a study on the subject from a group of engineers at Bell Labs. The report gave the investors confidence: The CATV industry was likely to be free from satellite competition for at least five years, or until 1964.

Predicting the impact of new technology on a nascent industry is a long shot in most cases. As it happened, the Bell Labs team was off by more than a decade when it came to timing the advent of satellite transmission of television signals to cable systems. And they had no way of knowing that when satellite technology arrived on the CATV scene—via a little-known service called Home Box Office—it would be one of the best things to ever happen to the business. That said, even under a worst-case scenario, the TelePrompTer team was confident it could earn a handsome profit on an investment in CATV, even if it had only five years to do so.

Hub Schlafly, who had worked on military radar applications during WWII and also held a patent on the TelePrompTer device, warned TelePrompTer's Irving Kahn and Monty Rifkin in the late 1950s about the future competition cable television would face from satellite technology.

Building a System

By January 1960, TelePrompTer owned the three systems in Rawlins, Silver City, and Farmington. The Schneiders backed out of doing a deal for the Casper station at the last minute. (They later bought out Daniels and other investors.) The station formed the basis for United Cable, the company they parlayed into a leading TelePrompTer systems operator. Once news spread of the company's newfound investment interest, the phones rang almost nonstop in the TelePrompTer Times Square offices. Soon it had acquired systems in Liberal, Kansas, and Elmira, New York. By the end of 1962, the company had 15 systems with more than 30,000 subscribers.

TelePrompTer's Irving Kahn demonstrated his company's commitment to its customers by visiting subscribers of TelePrompTer's system in northern Manhattan.

To help run the cable business in its early years, Rifkin reached down in the company ranks. He had noticed an ambitious young man, John "Jack" Gault, who was clearly overqualified for his job as head of the script department.[39] Gault started learning the operations side of cable television, and by 1961 he was sent to run the system in Elmira.[40]

The TelePrompTer system, with Gault at the helm, brought a flair for consumer marketing to Elmira that was missing from most cable systems during this period. It offered coupons and used appliance trade-ins to generate a sense of buzz about cable television in the community. Elmira chief engineer Austin "Shorty" Coryell worked closely with equipment manufacturers to develop one of the country's first 12-channel cable systems in the early 1960s, a time when many systems offered six channels at most. "The bottom line is people want to be identified with a winner," said Gault. "We were winners, and we were doing very good stuff."[41]

Daniels didn't do another deal with TelePrompTer after the first round of acquisitions, but he kept in touch with Rifkin. For the next few years, he cajoled Rifkin to come and join his company in Denver. When Rifkin suggested at one point that his wife's close family ties to New York, not to mention Rifkin's, were a major reason they were staying put, Daniels convinced them to visit Denver. They were wined and dined as his guests, marveling at the snow-capped peaks of the Front Range while lounging beside the swimming pool at the Cherry Creek Country Club. They were hooked, and relocated to Denver in 1963.

Bill Daniels, left, almost single-handedly made Denver the center of the U.S. cable industry by the early 1960s, attracting several other cable operators to the Mile High City, including Monty Rifkin, who joined Daniels' company in 1963.

Rifkin, president of the Daniels Management unit of the business, concentrated on the financial side of Daniels' operations—funding acquisitions and franchising—while Daniels focused primarily on his brokerage business. Gault followed Rifkin to Daniels and Associates in 1963, although he went to Pittsfield, Massachusetts, as Daniels and Associates' man running a system owned by investor Alfred Stern's Television Communications Corp (TVC), in which Daniels' company also held an equity interest. The following year he joined TVC and moved to Manhattan to serve as vice president of operations for TVC in Manhattan.[42]

Go-Go 1960s

As the 1960s progressed, cable antenna television was increasingly viewed by financial institutions and investors as a legitimate business that had made the transition from the mom-and-pop stage to that of an increasingly sophisticated industry. The threat of oppressive regulation and potentially ruinous competition remained a constant worry. But from a cash flow point of view, the industry had a lot going for it.

TelePrompTer hired Leonard Tow, who had a Ph.D. in economics and economic geography from Columbia University and had been acting as a consultant to the company, to serve as an assistant to Irving Kahn. The title was misleading: His job, following in Rifkin's footsteps, was to execute Kahn's strategy of expanding rapidly in community antenna television. But now Kahn was focusing the company's efforts on urban markets. To finance such an expensive proposition, compared to running cable down small-town alleyways, the TelePrompTer executives formed a handful of partnerships with Hughes Aircraft, which had a record of investing in promising new technologies, with separate partnerships focused on building out cable systems in portions of Los Angeles and Manhattan. Another Hughes partnership developed the potential for satellite transmission of television signals to cable systems. TelePrompTer was on a path of continued acquisitions and mergers that made it the nation's largest CATV system by the early 1970s, though during the same period scandals tarnished the company's genuine accomplishments in the field.[43]

One of TelePrompTer's largest targets was H&B. From a standing start in 1960, H&B American Corp. had amassed 90,000 subscribers by 1966, making it the country's largest CATV company at that time. H&B, a debt-laden manufacturing conglomerate headquartered in Beverly Hills, California, was attracted to the steady cash flow of CATV systems. In 1960, it paid $5 million for nine of Jerrold's CATV systems. By 1962, the publicly traded company had 22 systems with more than 63,000 subscribers. It was so committed to the CATV business that it sold the balance of its other business lines and concentrated solely on cable. H&B merged with TelePrompTer in 1970.

In 1963, Ralph Roberts bought the first cable system that formed the basis for the company renamed Comcast in 1969.

Alan Gerry put the Catskill Mountains of New York State on the cable industry map in 1963 when he formed Cablevision Industries (CVI) and rapidly became an industry innovation leader.

Bob Magness, who was in the cotton and cattle business in Texas in the early 1950s, got into community antenna television after learning about it from two installers whose truck had broken down and needed a lift. He built his first system in Memphis, Texas, in 1956 and, with Bill Daniels' help, founded a system in Bozeman, Montana, in 1958. He moved there and began importing stations from Salt Lake City using a microwave link. By the 1960s, Magness and a handful of business partners were borrowing heavily to buy or build systems around the country. His company, Community TV Inc., had full or partial ownership of 24 CATV systems, located mostly in the mountain states and Nebraska. He moved company headquarters to Denver in 1965 and in 1968 renamed it Tele-Communications, Inc. (TCI). By decade's end, his highly leveraged enterprise was the 10th largest CATV company in the country.[44]

One of the businessmen attracted to the industry during this period was Ralph Joel Roberts. He had sold his belt and suspender business in Philadelphia in 1961, worried that the increasingly popular sans-a-belt trend would severely limit future growth. Shortly thereafter, he met a friend who wanted to sell a community antenna system in Tupelo, Mississippi. Roberts didn't know anything about Tupelo or CATV, but he liked the steady cash flow attached to it. He bought the system and founded American Cable Systems in 1963. He changed the name to Comcast in 1969.[45]

Alan Gerry was fixing an amplifier on a rooftop television antenna one very cold February day in 1955 in the Catskill Mountains of New York. A salesman from Jerrold Electronics who was driving by saw him and sounded him out, over a cup of hot coffee, about building a cable television system in his community of Liberty, New York. Gerry, who had received electronics training in the U.S. Marine Corps and in technical school, had been selling and repairing TVs for a few years and had installed a number of antennas in town and connected clients by wire to them, but didn't really have a commercial system per se. Drawing upon Jerrold's knowledge of the equipment needed and the economics of the business, Gerry rounded up a handful of local investors and plunged in. When his system launched in 1956, he had 300 customers paying a one-time $130 installation fee and $3.50 a month to receive five channels.

The mountains of Pennsylvania had bragging rights when it came to claiming to be one of the birthplaces, if not the only birthplace of CATV, but Gerry was rapidly making the Catskills region synonymous with CATV innovation. By the mid-1960s, Gerry had bought out his rival CATV system owners in Sullivan County, usually for about $300 a subscriber, as well as his original investors. To expand into other regions, including in short order the suburbs of Boston, Gerry became one of the first systems operators in the Northeast to build a microwave distribution system that could tie together tens of thousands of subscribers. And his company, which changed its name from Liberty Video to Cablevision Industries (CVI) during this period, led most rivals in customer service, local origination of programming, and advertising in the 1960s.[46] Three decades later, Gerry sold his company to Time Warner.

Several major corporations, most with roots in broadcasting, publishing, communications, or technology, also built, expanded, or acquired CATV systems in the 1960s. Cox Communications, for instance, paid $1.8 million for a system in Aberdeen, Washington, in 1963, and the following year bought the system that Leonard Parsons had launched in Astoria, Oregon. By 1965, Cox, which owned five television stations and four radio stations, as well as newspapers, had full or partial ownership of CATV systems serving 10 communities and 40,000 subscribers, making it one of the larger multisystem operators at that time.[47] Divisions of CBS, NBC, General Electric, Newhouse, Westinghouse Broadcasting, Scripps-Howard Broadcasting, and Avco were among others actively building CATV businesses.[48]

By the late 1960s, some cable operators were offering 18 to 20 channels on their newly built or upgraded systems. One such system in Akron, Ohio, which became part of Warner Communications, demonstrated its breadth of selection by displaying each of its 18 channels on a separate television screen.

All Cable Service Was "Basic"

John Salinas
Head Technician (Retired 2010)
Corpus Christi, Texas

After more than 42 years in cable television, John Salinas retired in 2010 as head technician in Time Warner Cable's Corpus Christi, Texas, facility. When he got his first job in the industry in Beesville, Texas, in 1968, all cable service was "basic":

We were between San Antonio and Corpus Christi and we put a big 400-foot tower on a hill, and we were picking up five channels from San Antonio and three from Corpus Christi. That was cable. Then we got a weather channel, kind of a scanning temperature gauge or something like that, and a couple of other channels, FM user channels. That was the original cable system.

By 1968, in fact, broadcasters owned nearly one-third of all cable systems, and phone companies were expanding their hold on the industry as well.[49] The FCC responded by banning ownership of a cable system and broadcasting station in the same market, and forced the networks to divest their cable holdings, with the deadline eventually pushed back to 1975. In 1970, the FCC also effectively banned phone companies from owning cable companies in their service areas.[50]

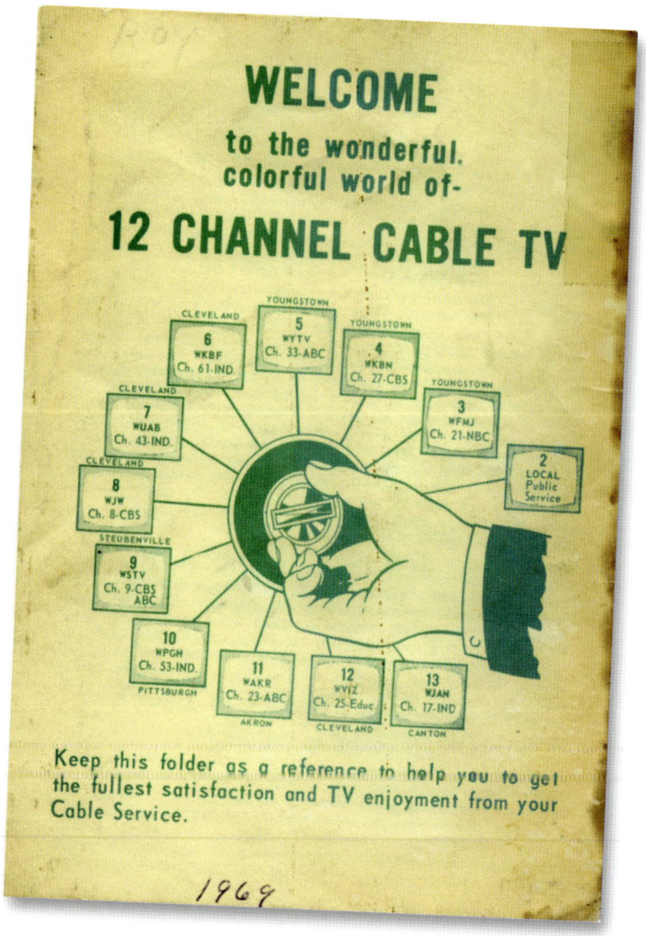

By the late 1960s, innovative cable systems such as this one in Salem, Ohio, offered an unprecedented 12 channels, all of them in color.

Time Inc.

Time Inc. reigned as one of the nation's premier magazine publishing companies throughout the 1960s. At its peak, the company accounted for roughly one-third of all magazine advertising spending in the United States. Its flagship *Time* magazine was a household name and brand throughout the country, and indeed around much of the world. Its authoritative voice, crafted by founder Henry Luce, in many ways spoke for America's surging prominence on the world stage as the nation emerged from the Great Depression of the 1930s.

Luce had personally coined the phrase *Pax Americana* to describe the United States' unquestioned dominance of global affairs as of the 1940s, in a none-too-subtle linguistic comparison with the Roman Empire. As the Soviet Union rose in power to challenge U.S. influence in the postwar era, *Time* rallied Cold Warriors around the world with its spirited and unrelenting attacks on this and other Communist systems, including revolutionary developing-world movements in Vietnam and elsewhere.

In terms of its corporate culture, Luce's entrepreneurial zeal had given way to modern management by the 1960s. Time Inc. was known for hiring very bright generalists, many with Ivy League, East Coast Establishment credentials. Landing a job at Time Inc. was considered a ticket to the good life, where teamwork and impeccable ethical standards were the norm, and executives worked and socialized together for years.

The Luce era may have been a thing of the past, but the company was still willing to take big risks on money-losing efforts for years, if it was convinced they had a shot at paying off. *Sports Illustrated*, for instance, lost money for more than a decade before finally showing a profit in 1966. By playing an instrumental role in the success of that magazine, its publisher, J. Richard Munro, was tagged as a rising star at Time Inc. and played an important role in the success of its cable television efforts in the 1970s.

During the 1960s, the company also began to think of itself as a communications company, not just a publisher of magazines and books. The rise of television made as much of an impact in the boardroom of Time Inc. as it did in the pages of the magazine. Indeed, by decade's end it was clear to management that television was one of the main reasons *Life Magazine*'s sales were slipping badly.

The company created Time-Life Broadcasting in the 1950s to diversify into broadcast television. It bought a number of stations from 1952 through 1957. The broadcasting subsidiary diversified into cable television in the mid-1960s, announcing in 1965 that it planned to build two systems in Michigan.[51] Cable was seen as another means of distributing its wealth of programming. Additional systems followed in Terre Haute, Indiana, and the northern portion of San Diego, California.

Experience it. Use your smartphone to watch a video of Time Warner Cable chairman and CEO Glenn Britt describing the origins of cable television.

Steve Ross' early years in the parking lot and limousine businesses gave little hint that he would become one of the most influential figures in the history of the cable television and entertainment industries.

Initially, the effort was something of an add-on to the broadcasting business. But by the latter part of the 1960s, Time-Life had systems or franchise rights in California, Indiana, Michigan, Texas, New Jersey, Pennsylvania, and New York City. Time Inc. sold off its broadcast stations located in San Diego and Bakersfield, California; Denver, Colorado; and Indianapolis, Indiana, to McGraw-Hill in the early 1970s, in part to offset losses from its floundering *Life Magazine*, and concentrated on cable.

In the late 1960s, Time Inc. took a 20 percent stake in a publicly held company called Sterling Communications, run by Chuck Dolan, which had the CATV franchise for the southern half of Manhattan. Time Inc. made additional capital investments in the enterprise, which, although on shaky financial footing for years, played a pivotal role in the company's expanding role in the cable industry beginning in the 1970s. Dolan, who, along with all other investors in Sterling, was bought out by Time Inc. in 1973, received Sterling's cable franchises on Long Island as part of his exit package, which he used to create what became CATV giant Cablevision.

Warner Communications

Warner Bros., among the leading motion picture studios in Hollywood, was definitely showing its age by the mid-1960s. Seven Arts Productions, a company that distributed films to television stations, bought the studio in 1967 to broaden the selection of films it could offer to the television industry. Within a few years, that combined operation caught the attention of an ambitious, enthusiastic, free-wheeling businessman determined to get into the entertainment business: Steven J. Ross.

Ross' early track record in business did not suggest he was headed to the top of one of the world's largest communications companies. He started out as an executive trainee at the funeral home run by his in-laws in New York City. He hit upon the idea to rent out the funeral company limousines after hours. Looking for a place to park his limos, he made a deal with a parking lot company, Kinney, and then merged Kinney with his company.

In 1967, Ross bought a private talent agency, Ashley Famous, whose owner, Ted Ashley, suggested he look at Warner Bros. for its movie vault, as well as a music subsidiary. Ross bought Warner-Seven Arts in 1969 and decided to sell off his non-entertainment businesses. As he looked for alternatives to distributing the studio's content, his focus within a few years landed on CATV.

Ross placed his bets on the entertainment and communications fields, and there was no turning back. Over the years, Warner demonstrated a keen ability to "anticipate new technologies and their resulting markets," as Ross told shareholders in 1987, reflecting on a quarter-century spent at the helm of a public corporation. Just as important as a nose for the next technological hit was a "willingness to build organizations to capitalize on those opportunities."[52]

Forming ATC

Systems operators that had grown up in the community antenna industry, keenly aware of the giant publicly held companies infringing on their turf, by the latter 1960s saw an opportunity to tap the public markets for a source of permanent equity capital for their companies. Wall Street was on a roll, and investment bankers were attracted to the steady cash flow they were confident would make CATV stocks an easy sell to the public. "Those of us in the industry, of course, were looking for another form of currency we could use to continue to develop and grow in this capital-intensive business," said Rifkin. "So we started kind of looking at the public markets."[53]

He took the issue to Daniels in 1968. "I sat down with Bill Daniels and suggested to Bill that this era of public companies was coming. That we had an opportunity to be a leader in it, and that this was something I had done before, I had been trained to do and that I'd love to do again.[54]

"So I concocted the idea of forming American Television and Communications Corp. (ATC), a holding company to in effect acquire by merger the 14 or 15 different entities that Bill and I and other investors owned, and then take that company, which would number about 45,000 subscribers at that point, public."[55]

Daniels thought it was a great idea. He even came up with the name. "He thought it would get us listed in the phone book right up there with American Telephone & Telegraph," Rifkin recalled with a chuckle.[56] But Daniels didn't want to personally be involved as an officer or director. "He didn't like the idea of living in a fishbowl." He did, however, like the idea of "securing liquidity for his equity interests" and being able to sell stock to raise cash as the need arose.

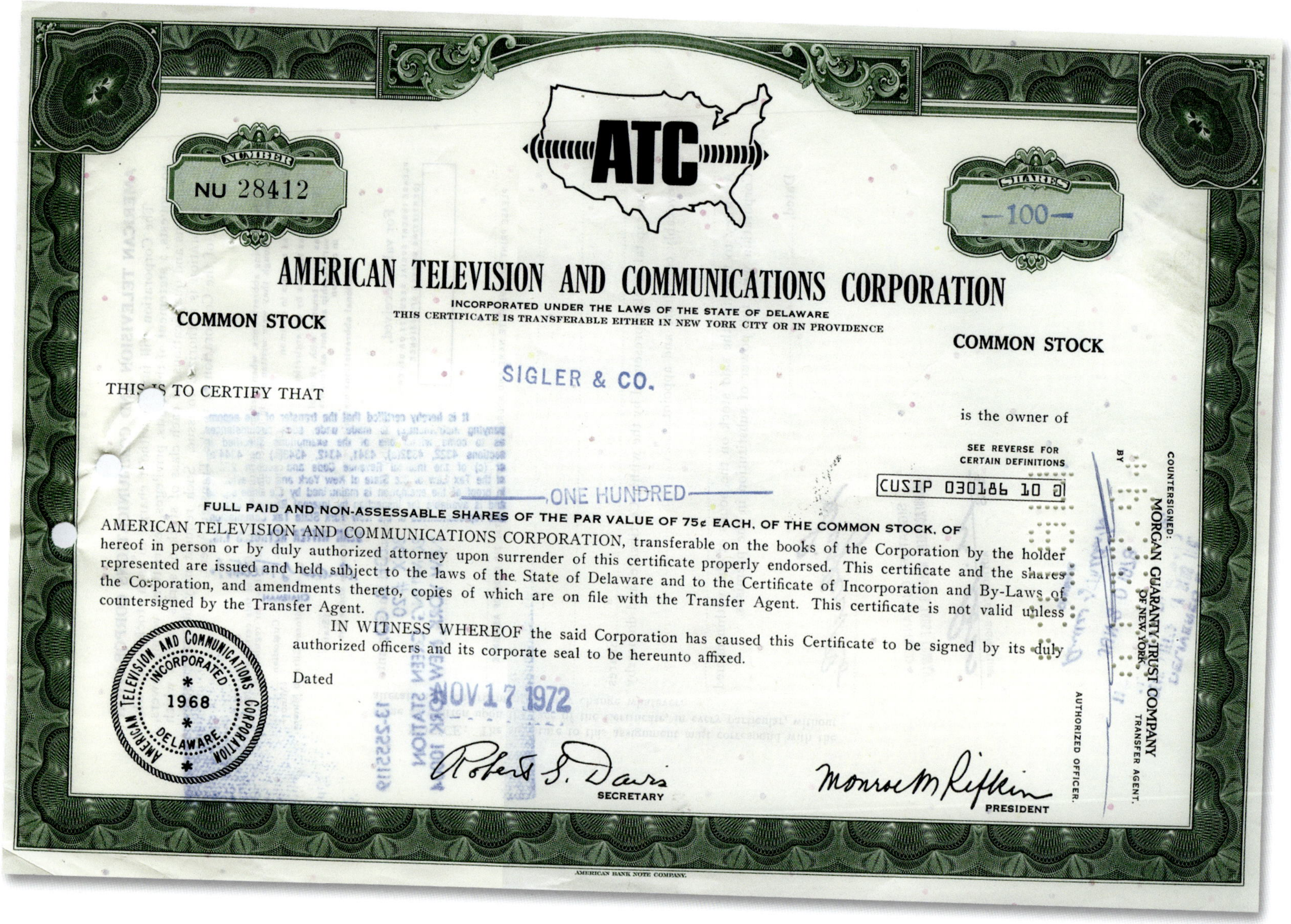

Monty Rifkin persuaded his partner, Bill Daniels, to combine cable systems they and others owned to form a public cable company, American Television and Communications Corp. (ATC). While several events over the past five decades represent major milestones in the formation of what has become Time Warner Cable, the company traces its roots to the June 4, 1968, founding of ATC.

Taking Stock

ATC, the cable company to which Time Warner Cable traces its operating roots, was formed on June 4, 1968. Rifkin worked with investment bankers from Paine Webber, Jackson & Curtis to structure the new company and prepare for an initial public offering of stock. The company was formed, as noted in the stock-offering prospectus, "to create a geographically diversified entity with centralized management and combined financial resources that could produce operating economies and serve as a medium for growth in the CATV industry through internal expansion and acquisitions."[57]

ATC, headquartered across the street from Daniels' company in 1,100 square feet of space, consisted of 36 CATV systems serving 65,500 customers in 14 states as of August 1968. The company, which had originally hoped to sell stock by yearend 1968, was buying systems at such a clip during late 1968 and in early 1969 that the offering was delayed so the Securities and Exchange Commission could review updated financial information. ATC had just over 100,000 customers by the time it went public in the spring of 1969.[58]

The stock offering raised $4.7 million to retire debt tied to the purchase of systems in 1968 and help fund the company's expected torrid growth. Monty Rifkin was its first president and the only corporate officer on its board of directors. He recruited Douglas Dittrick, who had been manager of operations of General Electric Cablevision Corp., to be vice president and treasurer.

Another call on ATC's capital during the 1960s was the need to reinvest in the continual technological advancements that characterized CATV during the decade. These included the shift by decade's end to transistorized amplifiers and the increased number of channels that systems could offer subscribers. As late as 1964, 85 percent of cable systems carried only three to seven channels.[59] With the Elmira, New York, system that Rifkin purchased for TelePrompTer in the early 1960s reportedly being the first in the country to go to 12 channels,[60] systems around the country followed. The next step was 23 channels, and by the late 1960s, systems were technically able to offer 36 channels, though few had made the investment to do so. That would come in the following decade. As Rifkin recalled, "Every one of those step-ups kind of created technological obsolescence and the need to upgrade and replace" at enormous cost to the system owners.[61]

The largest group of systems acquired through arms-length transactions to form ATC had been controlled by Narragansett Capital Corp., a small business investment company created and owned by Royal Little. The retired chairman of Textron and widely considered the father of American conglomerates, Little joined ATC as its chairman. Narragansett Capital, in return for contributing the largest number of systems to ATC, was the largest shareholder.

The next largest seller of systems to ATC, and second-largest shareholder, was the Memorial Drive Trust, a profit-sharing trust of the consulting firm Arthur D. Little. Systems were also purchased from the venture capital firm Boston Capital Corp, equipment manufacturer Spencer-Kennedy Laboratories Inc., and from Daniels and Rifkin, who jointly owned seven systems they sold to ATC. At the time of the stock offering, Daniels was the third-largest shareholder of ATC, and Rifkin the fifth-largest.

TOP
Monty Rifkin, left, recruited seasoned manager Douglas Dittrick from the cable television subsidiary of General Electric to help him run the newly formed ATC.

LEFT
Rifkin, known as one of the most efficient and reputable operators in the industry, positioned ATC for an era of explosive growth as the 1970s dawned.

The Cusp of Growth

A decade after pondering his pastrami sandwich at TelePrompTer, Monty Rifkin was ready to eat the cable industry's lunch. By the end of ATC's first fiscal year in June 1969, the company was preparing applications for franchises in a number of new cities with populations of more than 250,000. Even though cable industry subscriber growth was nearing 25 percent a year, there was still a lot of upside out there. Cable still only served about four million homes, or about 7 percent of the total number of television homes in the United States.

Obstacles remained, of course. The problem of copyright payments for programming carried on cable systems, despite the Supreme Court victory of 1968, wasn't likely to go away. And the FCC had created a thicket of regulations that had the effect of freezing most cable system expansion into the top 100 urban markets. The possibility of additional state and federal regulations, not to mention the demands of local authorities granting franchises, continued to hang over the industry as well.

The company was confident that these issues would be resolved. What the industry had to offer was just too compelling for regulations to stand in its way for long. Indeed, it is astounding how clearly Rifkin and Little, in their first letter to ATC shareholders dated September 16, 1969, could envision and articulate cable's potential, and the potential of the company that became Time Warner Cable. Making that potential a reality has been the challenge for company employees, then as now.

With cable, the TV set of the future will be more than a passive device; it will serve as a communications center capable of sending as well as receiving information.

We can foresee the morning newspaper being delivered by a new facsimile reproduction system; utility meters being read and signals sent back to a computerized central station; homes being equipped with fire detection and security systems designed to alert police and fire stations automatically; and merchandise displayed on the TV screen with orders being placed merely by pushing a button. The cable makes all of these possible.

In addition, we foresee the home of tomorrow linked, by means of the cable, with a high-speed computer, enabling instantaneous call-up and transmission of tens of thousands of different types and bits of information to the TV screen, adding tremendously to television's potential in the fields of education, instruction, professional training and adult education.

We believe that cable television systems, with their enormous channel capacities, are in an ideal position to become a leading force in helping to meet the expanding communications needs of a modern society. ATC has quickly proven itself to be a leader in this rapidly growing field, and intends to continue to be in the forefront of its growth.[62]

"We believe that cable television systems, with their enormous channel capacities, are in an ideal position to become a leading force in helping to meet the expanding communications needs of a modern society."

— Monty Rifkin and Royal Little, *1969 letter to ATC shareholders*

As American society prospered in the postwar decades, television became the focal point of many family gatherings. Cable television helped extend the medium's reach far beyond the cities with populations that could support one or more broadcast stations.

Chapter 2

Coming of Age

The 1970s

Public access programming, such as Teen Talk *in Brookline, Massachusetts, gave cable operators a way to differentiate programming in the marketplace and be responsive to community needs as the 1970s progressed.*

A relative latecomer to cable television in the early 1970s, Steve Ross quickly made up for lost time. His trademark enthusiasm and charisma were welcomed by systems operators looking to cash out the equity they had built up in their systems. Ross, focused as ever on the potential for growth in the industry he was interested in at the time, practically sold himself on the value of some of his early cable television acquisitions.

Even for an industry that was used to a fairly steady stream of acquisitions, Ross' buying spree stood out. In October 1971, Ross' company, Kinney National Services, announced the purchase of Television Communications Corp. (TVC), as well as Continental Telephone's cable properties. By May 1972, the renamed Warner Communications, Inc. also had agreed to acquire cable systems operator Cypress Communications, including the Pottsville, Pennsylvania, system that traced its roots to industry pioneer Martin Malarkey.[1] In less than a year, Ross had amassed cable systems serving more than 340,000 customers, ranking his company and ATC in close competition for second to TelePrompTer in number of subscribers.[2]

Ross' acquisitive reach may have exceeded his grasp of the cable business. Many of the systems he purchased were small-town systems with limited channel offerings that hadn't been upgraded in a decade or more. Systems in the Massachusetts towns of Medford and Somerville as of the mid-1970s, for instance, included such limited capacity amplifiers that two wires ran into the home: one for radio channels and one for a half-dozen or so television channels. Subscribers turned a switch to watch TV, or listen to the radio through their TV set speakers.[3]

At the same time, newly constructed 36-channel systems Ross purchased in Columbus and Akron, Ohio, were hugely expensive and didn't bring in much money.

Residents of these relatively large cities could receive multiple broadcast channels in their markets, and the cable systems weren't offering any additional programming of note. Customers were cancelling the service as fast as new subscribers were signing up.[4]

Gustave "Gus" Hauser, who at the time was exploring building his own telecommunications business, met Ross through legendary investment banker Felix Rohatyn of Lazard Freres. Hauser had been talking with Rohatyn about using his position as executive vice president of Western Union International to buy a controlling interest in that company. Instead, Ross convinced him to join Warner, naming Hauser CEO of the newly minted Warner Cable Communications division in 1973.[5]

TOP
ATC's status as a profitable, rapidly growing, publicly held cable television operator in the early 1970s soon attracted the interest of executives at Time Inc., whose cable and broadcast properties were not considered industry leaders.

RIGHT
Cable public access channels were well-suited for carrying programming featuring primary and secondary school events, such as spelling bees, as cable system operators worked to strengthen ties with local communities.

"The Next Big Thing"

Hauser had actually considered buying or investing in systems operator TCI in 1972, but the debt-ridden company, with its aging rural infrastructure, didn't look like the future of telecommunications to him. He was confident he could fix the problems at Warner's systems. But he wanted to invest in cable only if he had a shot at helping the medium realize its potential, much like Rifkin and Royal Little highlighted in ATC's first annual report. If not, why bother?

"I made a deal with Steve Ross that I would undertake to see if we could come to the next big thing in cable television," he said. Their goal was to "give people something ... that they wanted and see if we could build an industry out of this. And if we didn't, then we'd drop it."[6]

As the 1970s began, the cable television industry experienced explosive growth on all fronts. A new generation of business leaders and investors recognized cable's largely untapped potential and flocked to the industry. And by mid-decade, satellite transmission of television signals exclusively for cable systems marked the dawn of a new era in cable television technology. Cable for the first time had a truly national reach, and that sparked demand for cable programming to attract customers in urban America, CATV's new frontier by the late 1970s and early 1980s.

ATC Growth

Ross' initial investments in cable may have put Warner Cable nose to nose with ATC in the industry rankings, but that didn't mean Rifkin was sitting idly. Quite the contrary. ATC's acquisition binge, which had delayed its initial stock offering in 1969—in fact, the company had received the National Venture Capital Award in 1969 for its "skillful appraisal of cable television growth potential"[7]—gathered momentum in the 1970s.

Rifkin, while keeping a tight rein on spending, saw nothing but upside for the industry. Meantime, ATC's expansion caught the eye of another cable operator, Time Inc., which was dissatisfied with the progress of its own cable and broadcast businesses. Before the end of the decade, Time Inc. bought the young company and made it the core of its television operations.

People who knew ATC mainly through Bill Daniels' pivotal role in putting together the systems that formed the company, and his substantial stock holdings, didn't have a full appreciation of Rifkin's role in ATC's subsequent success. It didn't help that Rifkin lacked Daniels' back-slapping and speech-making skills. But to those who worked closely with Rifkin, there was no question as to who was in charge. David O'Hayre worked initially as an outside auditor of ATC's books before joining the finance department in 1973, rising to controller by the late 1970s. As he recalled, it was Rifkin's way or the highway:

He was as tough as anybody I ever knew. He was definitely in charge, ask anybody. It was his company. He had his hand on every nickel. He knew where everything was going, what was being spent, and not just on the financial side. He knew how the operations were going. He was the boss. He had a difficult time working with a lot of people, but I always got along fine with him, I guess because we were both CPAs and had the same outlook toward business: Don't spend a nickel extra.[8]

For years, Rifkin approved every spending request that came through the Denver office, and would often approve only a fraction of what was requested. As Jack Gault recalled, "If you wanted 50 rolls of toilet paper you requested 100, because you knew every request was going to get cut in half."[9]

That same approach even applied to hiring in some cases. Barry Rosenblum joined ATC in 1979, working in sales for one of the area systems that was partially backed by a group of local investors. "I came on at $18,000 a year, and I actually had to get two separate paychecks because Monty wouldn't approve anything more than $14,000," he said. "So the local group of investors had to pay me $4,000."[10]

Barry Rosenblum joined the ATC system in Jacksonville in 1979 and had firsthand experience with Monty Rifkin's tight controls over expenses. Local Florida investors in the system paid a portion of Rosenblum's salary above that offered by Rifkin in order to attract the young sales executive to ATC. Rosenblum would go on to manage some of Time Warner Cable's largest operations, including New York City, Los Angeles, and Texas. In 2010, he became the company's senior vice president of news and local programming.

Pete Conrad: The Right Stuff

Astronaut Charles "Pete" Conrad Jr. served on ATC's inaugural board of directors and was the company's chief operating officer from 1973 to 1976.

ATC's legendary co-founder Bill Daniels surrounded himself with some equally colorful characters over the years. None more so than Charles "Pete" Conrad, Jr., who served on ATC's inaugural board of directors, and as ATC's chief operating officer, from 1973 to 1976.

Conrad, who met Daniels as a precision flyer for the U.S. Navy, went on to achieve worldwide fame as a U.S. astronaut and the third man to walk on the moon. He was featured prominently in *The Right Stuff*, Tom Wolfe's portrait of the Mercury space program (even though Conrad didn't fly in space until he served as an astronaut in the Gemini space program); and in *Rocketman: Astronaut Pete Conrad's Incredible Ride to the Moon and Beyond*, written by his second wife, Nancy Conrad, and Howard Klausner. On his final mission in space, Conrad used brute force during a spacewalk to free a solar panel on Skylab 2, thereby salvaging the mission.

Conrad raised ATC's national profile with regulators and legislators in Washington, and he was a magnet at industry and client meetings. Standing a mere five-and-a-half feet tall, it was nonetheless easy to find Conrad in a crowd, recalled Jimmy Doolittle, who had joined ATC as a system manager a few years earlier than Conrad, because Conrad was invariably in the center of it. "You walk into a room and everybody's going to huddle around Pete," Doolittle said, "because Pete was the best storyteller you ever heard. And he could talk your head off."[11]

Returning to his aviation and aerospace roots, Conrad joined McDonnell Douglas in 1976. Reuniting with Daniels in 1996, three years before his death as a result of injuries suffered in a motorcycle accident, Conrad was a crew member on a Lear Jet owned by Daniels that set a record for an around-the-world flight.

"This Is a People Business"

Rifkin, setting a standard that has endured to the present day, also placed great importance on attracting and developing talented executives and managers at all levels of the rapidly growing company. While most of his rivals were still promoting technicians who may have been splicing cable in the 1960s to managers in the 1970s, Rifkin was knocking on business school and industry doors to attract the best and brightest. At the same time, he was intent on developing the skill levels of his existing technical staff. In 1970, at company headquarters in Denver, he created one of the industry's first full-fledged engineering departments to design and coordinate the construction and maintenance of ATC's systems and franchises.[12]

As Rifkin, then 41, told the *Rocky Mountain News* in May 1971, "The key to our progress is our management team. This is a people business, and the people on our staff are the most valuable resource we have. The most important aspect of this business is what kind of creative talent you have got in management."[13]

Following his recruiting of Doug Dittrick from General Electric Cablevision Corp. on the eve of taking ATC public, Rifkin over the next few years added former FCC attorney Bruce E. Lovett as ATC's vice president for corporate development and its eyes and ears in Washington, D.C. Lovett also served as general counsel for the NCTA, the industry's main trade group and lobby arm, during this period. Another FCC staffer, David Kinley, joined ATC in 1976 and played an important role in helping the company gain urban franchises around the country.[14]

Rifkin also brought back into the ATC fold Jack Gault, the fellow TelePrompTer alum who had followed him to Daniels and Associates and then ended up as president of Continental CATV Inc. and Commonwealth Television Inc. in New York. In 1971, Gault was named ATC's vice president of marketing.[15]

Rifkin hired John McDonough from Litton Industries in 1971 as vice president of finance. The hard-charging executive played an important role in major transactions, including ATC's purchase of Jefferson-Carolina Corp. and the purchase of Time Inc.'s cable systems in exchange for stock. The following year, Michael McCrudden joined ATC after receiving a master's in business administration from the University of Pennsylvania's Wharton School. He rose rapidly through the operating ranks, serving as senior vice president, Western operations, by decade's end.[16]

In 1972, Rifkin recruited Joe Collins right out of Harvard Business School, where he had become intrigued with the industry's potential after a prospective buyer asked him to write a report on cable television's prospects. Two years later, Rifkin, anxious to instill some marketing savvy in the corporate DNA, recruited Trygve Myhren, a veteran marketing executive with an MBA from Dartmouth who had learned his trade at Procter & Gamble, to join ATC, which he did a few months later in the spring of 1975. Myhren and Collins rapidly rose to the top echelon of ATC's executive ranks and eventually served as the company's second and third CEOs, respectively.

Rifkin led the cable industry in attracting executives from outside the business, such as Procter & Gamble marketing executive Trygve Myhren, left, to help ATC extend its reach and attract new customers. Myhren later succeeded Rifkin as chairman and CEO of ATC.

"One of the really big, major contributions that Monty and ATC made to the industry was bringing in a lot of sophisticated people, at least by training and a lot of cases by execution, into a business that hadn't had that before," Collins later noted. "They were able to join a lot of people who were also very entrepreneurial, and that made for a very good mix for a developing company and a new and growing business."[17]

Another of Rifkin's most influential hires actually started with the company in 1969 as his secretary and the company's 13th employee. "You sort of did everything that came across your desk," June Travis recalled. "There were no departments. It was very entrepreneur-ish in those days."[18] Travis rose rapidly through the ranks at ATC's Denver headquarters, while also getting an executive MBA from Denver University. She was named director of communications, and later vice president in administration and operations, and then executive vice president of operations.[19] Travis also became a role model for women entering the CATV industry, co-founding Women in Cable (now Women in Cable Telecommunications) in 1979 and becoming a senior executive at the NCTA.

One of the most influential women in the early decades of the cable industry, June Travis joined ATC in 1969 as Rifkin's secretary and rose through the ranks to become executive vice president of operations. In 1979, she co-founded Women in Cable, and later became a senior officer of the NCTA.

Experience it. Use your smartphone to watch a video of former Time Warner Cable CEO Joe Collins describing Monty Rifkin's farsighted hiring practices

Judy Braden, dispatch supervisor in Kansas City, with her mother, Joyce Bramble (who retired in 2003 after 23 years of service); and sister Kim Holsted, advanced cable store specialist. Judy celebrated her 25th anniversary with Time Warner Cable in July 2011.

"We All Care About Each Other"

Cable television was a family affair in many smaller communities around the country. In the late 1970s, Joyce Bramble, who handled most of the paperwork for ATC's small American Cablevision of Kansas City office as secretary to the president, would bring employee time sheets home to do the office payroll. Her teenaged daughter, Judy, diligently helped her mother alphabetize the sheets on the kitchen table.

"It was just a real small family kind of atmosphere at that point," said Judy, who started working for the company herself in 1986 as a dispatcher on the 3 p.m. to midnight shift. Today, Judy Braden is dispatch supervisor and has worked in numerous positions over the past 25 years. But as much as the technology has become more sophisticated and the size of Time Warner Cable's presence in Kansas City has grown, she credits the company with nurturing a family culture in the workplace.

"We all care about each other. It's a family. Even as large as the company has gotten, all the coworkers love each other and everybody gets along like it's a big family. Sure every once in a while you don't get along just like a normal family, but everybody cares about each other, and they care about the customers and so that has made the company a success."

Building new cable television systems in the 1970s, especially in urban areas, required a well-trained workforce employing sophisticated installation techniques.

Pole Farms, Growing Complexity

On-the-job safety concerns and the growing complexity of cable television technology spurred a number of technical training programs in the 1970s. Learning at the elbow of a more experienced technician remained a key element in any beginner's routine, but it would be supplemented with training courses. ATC constructed a "pole farm" near its Denver headquarters so new hires could gain experience under close supervision and favorable conditions. Warner Cable had similar facilities in Malden and Medford, Massachusetts.[20] "We had a pole farm so that they learned how to climb safely," said ATC's June Travis. "That was probably our very first venture into anything that would have looked like formalized training."[21]

Travis noted that the growing complexity of industry technology in the 1970s created demand for additional training. "When I joined the company, we were running five channel cable systems; that was the standard," said Travis. "Once we got beyond 12 [channels], which was sort of mind boggling, you needed a converter, which sounds really great except that a technician didn't necessarily understand what a converter was," she said, referring to the boxes that were installed in customers' homes so that the additional channels could be viewed on their existing televisions. "We started working with technicians ... to keep them abreast of a very rapidly evolving technological part of our business."[22]

As the cable industry continued to expand during the 1970s and cable technology became more complex, ATC and other operators instituted training programs to boost technicians' and installers' skill levels.

Buy or Build?

In 1970, ATC acquired cable systems in Albany, New York, its largest market to date, as well as Reading, Pennsylvania, and Jackson, Mississippi. Smaller systems were added in North Carolina, Wisconsin, and Oklahoma. A number of other smaller systems were added the following year, including in five counties surrounding Albany, as well as in West Virginia and Ohio.

While it continued to buy existing systems, ATC recognized that its future lay in acquiring franchises to build new systems in larger and faster-growing markets. "In the past, ATC has grown principally through the acquisition of existing CATV systems," the company stated in a November 1971 Securities and Exchange Commission filing for a second offering of common stock.[23] "But management believes that development of new CATV franchises will become a more significant factor in achieving future growth."[24] As Rifkin later recalled, "We decided this acquisition stuff was OK, but when you made these acquisitions, you were buying technologically semi-obsolete properties in limited growth" markets.[25]

That wasn't to suggest that the company didn't recognize its obligation to service existing customers in smaller legacy systems. By the early 1970s, for instance, ATC owned the system in Franklin, Pennsylvania, started by the Harter brothers that, as noted in Chapter 1, was one of the oldest in the country. Collins said, "If you went into downtown Franklin and you put up an antenna you would get nothing, so we virtually had every house in Franklin as a customer, and there was no marketing. Everybody had it. And in fact, because Franklin had some economic difficulties, you'd disconnect your telephone long before you'd disconnect your cable. ..."[26]

Rifkin followed Walt Disney into the Orlando, Florida-area market in 1970, getting in on the early stages of growth that would characterize the central Florida city for decades to come.

Orlando: "My Pride and Joy"

Rifkin got the idea for one of the most significant franchise acquisitions in ATC's history while taking a brief vacation in Florida in 1970. He read an item in a local paper indicating that the Walt Disney Company was contemplating building a major amusement park in the Orlando area. "I think Orlando was served by three television stations at that point. So I got the idea of our seeking a franchise. Thought it would be one of the great markets if Disney ever did come. Orlando was a sleepy town but larger than those that cable was typically in even at that time."[27]

ATC won the Orlando franchise and spent the next few years acquiring franchises in surrounding communities in order to serve the broadly defined Orlando market. The company announced plans in 1970 to build a microwave network connecting 18 towns and three counties in central Florida.[28] "I sent Joe Collins down to Orlando to manage it and sent Jack Gault running up and back down there to work on the surrounding franchises," said Rifkin. "The city of Orlando was only a small piece of that whole market. It took, I think, 10 additional franchises to really cement up the population and give them what they've got today. So we were doing great things, we were growing ... and we were very visible." Orlando, Rifkin added, "was my pride and joy."[29]

"It Was So Easy to Pay Your Bill, a Dog Could Do It"

Joe Collins recalled a common sight at the Reading system during the early 1970s that underscored the "basic" in basic cable systems of this period: "The office was in an old milk plant in downtown Reading ... and you would come through this front door and here were these white tiles and this counter. And so there was a customer, and once a month the customer had trained his dog, and the dog would come into the office with the coupon and $5, and it would go up on the counter and give the lady the coupon. She'd give the dog back the receipt, and the dog would go down and leave the store. And it was so easy to pay your bill, a dog could do it."[30]

Technology Test Bed

The state-of-the-art Orlando system was also an early test bed for gauging how quickly and realistically the dream of cable's two-way, broadband future might be realized. ATC engineers worked with equipment suppliers to develop one of the first two-way systems that could be capable of delivering movies, plus banking and shopping services, among others. Investors interested in the industry and systems operators from around the country flocked to Orlando to view a demonstration of these cutting-edge capabilities.

After a few years, ATC had to admit that the system wasn't ready for prime time. "At the end of the day, the technology really wasn't there yet," Collins said. "There was a problem with the upstream 'noise' on the technical side, and the boxes themselves were about as big as a suitcase, and it sort of sat behind the television set."[31]

Company engineers, as well as industry suppliers, learned a lot about the challenges still facing the industry. ATC took these lessons learned, and Rifkin in particular made a point as the decade progressed of not over-promising what cable television could deliver, even when confronted a few years later with a marketing blitz from rival Warner Cable that would claim to set a new standard for CATV technology and customer offerings.

Wired Nation

By the late 1960s and early 1970s, there was a growing awareness of cable television's impact, and its potential for meeting multiple urban needs. On the one hand, it was seen as a potential gold mine—or in the words of New York City Mayor John Lindsay, an "urban oil well"[32]—for the franchise fees it could produce for cities. More broadly speaking, CATV was seen as the ideal medium for creating the telecommunications network of the future for urban America. And it was considered particularly well-suited to serving the needs of, and giving voice to, America's various ethnic groups.

ATC's executive team of Monty Rifkin (seated) and from left, Jack Gault, Bruce Lovett, John J. McDonough, and Douglas Dittrick was among the most aggressive in the industry in the early to mid-1970s as they acquired or built systems across the country, including a string of suburban systems surrounding Orlando, Florida.

"It Was Easy to Get Ahead"

Ambition, work ethic, skill set. Key ingredients for success in the cable industry of the 1970s, and the same holds true today.

Gerry Campbell joined Warner Cable in Somerville, Massachusetts, in the 1970s as an ambitious twenty-something from the Boston area looking for work. He quickly discovered the company was a great launching pad for a career in the cable industry. "I went through general managers once every year—probably the average was nine months—because they couldn't find people. One of the things when I reflect on it, or when people talk about what you did, is this was just a boom time, so if you had any kind of technical skills, and leadership skills, it was easy to get ahead. And you just watched others fail too, so you had an opportunity to see what you should do right and what [not] to do wrong.[33] I was originally in service, and then over time, I went into project management." Within a few years, Campbell became the vice president of operations for the Medford complex, and today is the company's executive vice president in charge of Business Services.[34]

Report followed report, raising the stakes for policymakers' and the public's expectations of what cable could, and should, deliver. An influential report commissioned by Lindsay and released in September 1968 recommended a state-of-the-art 18-channel system for New York City's five boroughs.[35] A December 1968 report from Washington, dubbed the Rostow Report after its principal instigator, Undersecretary of State Eugene Rostow, declared cable TV's multichannel capacity the ideal telecommunications medium of the future. It also highlighted the promise of satellite distribution of programming to cable systems.[36]

An influential industry report issued in October 1969 by the Industrial Electronics Division of the Electronics Industry Association called for the sort of national investment in cable that the interstate highway system was receiving. It also looked more than a decade ahead and offered an eerily prescient take on the ultimate impact of broadband communications. This report was highlighted in an influential publication, "The Wired Nation," which first appeared as a special issue of *The Nation* on May 18, 1970, and was issued in an updated version as a book by the same name in 1972:

The Industrial Electronics Division, Electronic Industries Association (IED/EIA) views the services to be provided by broad-band communication networks in the late seventies and early eighties of landmark importance. We look upon such systems as being of "national resource" dimensions and the development of these resources as a national goal… The mushrooming growth in available information and the demand for access to this information is bringing about a revolution in communications which will produce a profound change in the way society is structured and the way we live.[37]

Urban leaders, led by New York City mayor John Lindsay, saw cable systems as vehicles that could serve multiple entertainment and educational needs by the early 1970s. Lindsay, eyeing the monthly revenue streams generated by cable systems that could be shared with city governments, compared a cable system to an "urban oil well."

"An Impressive Report Card"

In his first few years as CEO of Warner Cable, Hauser—even while forming a team to create "the next big thing in cable"—addressed the company's operating problems while also adding to Warner Cable's subscriber base. On September 30, 1976, Hauser took the podium at a meeting of the New York CATV Analysts Group to bring Wall Street up to date on Warner Cable's progress. It was an impressive report card.

After three years of operation, Warner Cable increased its subscriber base by more than 20 percent to in excess of 550,000. Expenses had been reined in every year even as the number of subscribers was growing. That helped drive a nearly fivefold increase in operating income (revenue less operating expenses, before depreciation and amortization) from $4.7 million in 1973 to roughly $22 million for 1976. And the company's pay-per-view movie channel, Star Channel, delivered at this point by local systems playing one-inch tapes at their headend facilities, was being subscribed to by 25 percent of the Warner Cable customers to whom it was offered.[38]

Warner Cable's studios in Columbus, Ohio, were the center of cable programming innovation by the late 1970s.

Many early cable television programming efforts focused on current affairs and included officials and speakers from local markets.

Public Access

In the industry's early years, community antenna television systems had intentionally emphasized their role as mere collectors and retransmitters of broadcast signals. Broadcast's supporters on Capitol Hill and at the FCC warned of the programming threat from cable for years. What modest programming that did exist tended to be camera scans of weather instruments. Modest public affairs programs and coverage of local school events were also featured on some newer systems.

The FCC reversed course to an extent in the 1970s and started championing cable television programming, though its enthusiasm was limited mainly to public access programming. Rules promulgated in 1972 and touching on nearly every industry issue—and running to more than 300 pages—required that systems with more than 3,500 subscribers had to offer at least one access channel. Systems in the top 100 markets had to provide three public access channels for public, government, and educational use, and a fourth that could be leased commercially. In return, compromises were reached or at least promised on other contentious issues, such as copyright payments.[39]

Cue the Gecko

Cable operators also experimented with advertising associated with local origination programming. Weather wheels displayed ads on revolving wheels or drums following a series of camera shots of a thermometer, barometer, and other gauges. Some advertising efforts were more rudimentary than others. Jim Chiddix was a technical jack-of-all-trades for a small cable operator on the island of Oahu in Hawaii in 1971. After a few hours of local programming during prime time—school-age hula competitions were a sure hit—the camera focused on a collection of display ads:

It was sort of like a weather wheel except that we couldn't afford one of those, so we found a jewelry display case, the kind that used to go around to show jewelry. We rigged it up so it would advance one step at a time and then hold there for 30 seconds, and then advance another step, and on each of these little shelves we'd put a 3 x 5 card and shine a camera on it. That was in a little shed out by the headend. Now Hawaii has a wonderful climate, but there's a lot of insect life in Hawaii, and other kinds of local wildlife, and you just learn to live with cockroaches running around and little lizards and stuff. We had bright lights shining on these 3 x 5 cards as they came around—and part of the entertainment if you watched it for a while, especially in the evening, moths would come and they'd sort of light on these cards because of the lights. Once in a while, if you were lucky you'd see a big lizard dart out and nail a moth and drag it off. So a little blood sport there, on cable. But we would charge people, I think five dollars a week or something, to put classified ads on the 3 x 5 cards as they went around.[40]

Jim Chiddix, who later helped lead ATC and the cable industry into the era of fiber-optic broadband communications, got his start in the industry with a small operator on the island of Oahu in Hawaii.

Jimmy Doolittle joined ATC as it expanded into the Carolinas in the early 1970s and later served as president and chief operating officer of the cable company for a decade after headquarters moved from Denver to Stamford, Connecticut, in 1988.

A former NASA rocket scientist, Jim Cottingham rose through the operational ranks at ATC and Time Warner Cable during his 29-year career, serving as the first president of ATC's national division in the mid-1980s, and later as executive vice president when the company moved to Stamford.

ATC-Cox Merger?

The heightened potential for CATV's expansion into larger markets by the early 1970s drove cable television stock prices higher on Wall Street even though a scandal rocked the industry's largest player. TelePrompTer CEO Irving Kahn was found guilty in 1971 of bribing three officials in Johnstown, Pennsylvania, tied to the company's 1966 winning of the CATV franchise there, and served time in federal prison. While not denying making the payments of $5,000 each, Kahn maintained he was the victim of an extortion scheme.[41]

Cable executives were certainly happy to see their stock prices rising. Since most spent more time on Main Street than Wall Street, however, they were increasingly concerned by the rising costs associated with building out urban franchises, which appeared likely to be 10 times that of rural franchises, or more. Many went to the equity or debt markets to raise additional funds. Others sought out business partners to share the load.

ATC, which raised $23.5 million from a consortium of insurers in 1972, also reached out to Cox Cable Communications, which was roughly the same size as ATC at about 250,000 subscribers, to discuss a merger. The two made a good fit strategically and, by pooling their resources, could more easily shoulder the expected costs of urban expansion. Even though the Justice Department had allowed the merger of much larger TelePrompTer and H&B two years earlier, the authorities had experienced an apparent change of heart about concentration in the cable industry by 1972. They rejected the merger outright in 1973.[42]

Rifkin considered suing to get the ruling overturned but realized that the two companies would be operating in limbo as the process played out. Municipalities could argue they didn't know whom they were issuing franchises to—the combined entity, Cox, or ATC?[43] The companies decided not to contest the ruling.

Shortly before it tried to merge with Cox, ATC agreed to acquire Jefferson-Carolina Corp., based in Charlotte, North Carolina. It was half-owned by Jefferson-Pilot Insurance Co. and United Utilities, which owned Carolina Telephone. Once the FCC had ruled that phone companies couldn't own cable systems in their home markets, the two decided to put Jefferson-Carolina on the market.

James "Jimmy" Doolittle had helped build the Jefferson-Carolina system from the ground up. He managed the first system, in Jefferson, North Carolina, starting with just 250 subscribers. Progressively larger systems followed. When he learned it was going to be sold, he decided to join ATC as system manager in Fayetteville, North Carolina. After two years, he was made regional manager of all ATC operations in North Carolina. Several years later Doolittle's career path would lead to president and chief operating officer of ATC.[44]

Time Inc. and ATC: "They Wanted to Hitch Their Wagon to Our Star."

Rifkin didn't have too long to brood after the collapse of the Cox deal. In 1973, he received a call from Morgan Guaranty, which he had retained as ATC's bank when it went public, hoping that its storied reputation would rub off on his small company. That same thought process on Rifkin's part played an important part involving another Morgan client—Time Inc.—on whose behalf the Morgan banker was calling.

Time Inc. had sold its half-dozen broadcast television properties, some dating to the 1950s,[45] to McGraw Hill Publishing in 1971, reasoning that cable offered the better growth potential. The company also needed additional funds to shore up its faltering *Life Magazine*, the Time Inc. property suffering the most as consumers, and advertisers, shifted their attention from picture magazines to television. But Time Inc. executives admitted that they lacked in-house management expertise when it came to television of any sort. (Rumor had it that some among senior management still rued the day that founder Henry Luce had reportedly passed on an opportunity to invest in ABC when the so-called third network was in its infancy.)[46] Since the mid-1960s, Time Inc. had acquired interests in 11 CATV franchises or systems serving 60,000 subscribers, from San Diego, the largest system, to Terre Haute, Indiana.[47] "They were not pleased with the progress they were making," Rifkin said. "In their words, they wanted to hitch their wagon to our star. We were delighted with that, so we entered into negotiations."[48]

"We didn't know the business," conceded Richard "Dick" Munro, the head of Time Inc.'s video group during much of the 1970s and one of the executives who met with Rifkin. "But we knew enough about it to know that this was a very highly respected cable company and one we'd be proud to be a part of. And that's when we made the initial investment," he added.[49] ATC used its own stock to buy a number of systems from Time Inc., which resulted in Time Inc. owning 9 percent of ATC.[50]

To help strengthen the bond between the two companies, Rifkin invited Time Inc. president and veteran journalist James "Brass Knuckles" Shepley to join the ATC board. "At first he thought that we were too small and he wouldn't waste time with that," said Rifkin. "I twisted his arm, and Jim came on our board. I think it was a great move for us. He was a great board member, a very pragmatic, tell-it-like-it-is, get-right-down-to-action kind of guy. Gave us further credibility out in the financial community and throughout the industry. Started our relationship with Time Inc."[51]

Throughout much of the 1970s, ATC played a significant role in the growth of Time Inc.'s video group, as indicated by this image from the 1979 annual report.

"A Huge Mistake"

One of Time Inc.'s key cable systems was not included in the ATC transaction. Time Inc. had desperately wanted Rifkin to take over its money-losing Sterling Manhattan Cable system as part of the 1973 transaction. (The name was subsequently changed to Manhattan Cable.) But Rifkin, who prided himself on producing earnings gains for his public shareholders every quarter, said he would rather walk away from the deal than add Sterling Manhattan to the mix. The sky-high cost of building in Manhattan, and myriad problems related to operating in such a high-density urban market, were just too much. A struggling programming subsidiary called Home Box Office was hardly an added inducement.

Sterling Manhattan struggled to expand its system in the southern half of Manhattan during the late 1960s and early 1970s.

"We were concerned that the problems there were huge. That growing out of those problems would have been an anchor around our string of successes and growth pattern," Rifkin said.[52] He later acknowledged that, with the benefit of hindsight, the decision was "a huge mistake."[53]

At the time, many within Time Inc. and without would have had trouble disputing Rifkin's logic. David Van Valkenburg, at the time a securities analyst following cable television stocks for the mutual fund group Investors Diversified Services (IDS), knew Rifkin well. He counseled the ATC president in no uncertain terms that ATC, with nearly all its systems in relatively small cities and towns, should steer clear of the troubled urban system. Rifkin followed his advice, and in the fall of 1973, he also hired the analyst to work in operations at ATC, initially on the West Coast. By the late 1970s, Van Valkenburg was vice president of Western operations.[54]

Sterling had been operating the cable franchise for the southern half of Manhattan since the late 1960s, and TelePrompTer the northern portion of the island.[55] Time Inc. had initially provided financing to owner Chuck Dolan. With losses mounting, Time Inc. bought out Dolan and other investors in 1973. As part of his separation from Sterling Manhattan, Dolan received cable franchises on Long Island. He used them to launch his next enterprise—Cablevision.

The Time Inc. management team that took over Manhattan Cable in 1973 worked overtime to iron out differences with local unions and city officials in order to control costs and push through a monthly rate increase, leading to a profit for the loss-plagued business after only about two years.

Even for Time Inc., which was known for being willing to stick with a business through tough times if there was a prospect of future success, Sterling Manhattan was looking like a lost cause. Time Inc. reported a $10 million loss on the southern Manhattan system in 1973 and had invested a total of $45 million in Sterling as of yearend 1973.[56] As Munro recalled, "We tried desperately to extricate ourselves from Sterling Manhattan."[57]

Munro thought at one point that he might have a buyer for the system in Warner Cable. Steve Ross had acknowledged that the business had problems but gave an indication that he might be sold based on the potential for a system in the heart of the Manhattan market. Further due diligence curbed his interest. "Warner came and took a look and saw what a dog it was," Munro said, and the deal was off.[58]

"There was nothing we could do," said Munro, "except try to dig ourselves out of the hole—to bring in new management and try to make Manhattan work."

It was an arduous process. Righting Manhattan Cable also would be a proving ground for the next generation of corporate leadership. Richard Galkin, a broadcasting veteran, initially was brought in to run Manhattan Cable. Wanting a tighter rein on the business, Time Inc. management in short order tapped Nick Nicholas from Time Inc.'s finance department to take charge of Manhattan Cable. Thayer Bigelow, also from finance, served as Nicholas' deputy. Gerald Levin, a young attorney who had been negotiating entertainment deals for Home Box Office, was put in charge of that unit.[59]

Demo vans that crisscrossed the city enabled New Yorkers to witness the improved image quality possible with cable television without having to travel to Manhattan Cable facilities.

Save It or Sell It

One day in 1974 after he had been assigned to Manhattan Cable, Nick Nicholas ran into Glenn Britt, with whom he had worked in finance, in the hall of Time Inc.'s Midtown offices. Britt, who had been hired two years earlier by Time Inc. out of Dartmouth's Amos Tuck School of Management and was one of only three MBAs hired that year, said, "If you need some help down there, here I am." Nicholas didn't waste any time. "The next thing I knew I was the finance guy in Manhattan Cable," Britt said.[60]

Britt quickly began to question the career move. He left the storied Time Inc. offices and moved to Manhattan Cable's offices on 23rd Street, at the time not one of Manhattan's nicer neighborhoods. In addition, it was not at all clear that the parent company considered Manhattan Cable a long-term part of the family. "My perception was we were there to make it a little better, and then somehow dispose of it," Britt said.[61]

Under the leadership of Nicholas and Bigelow, the team pitched in and turned a loss-plagued operation into a moneymaker in about two years' time. They took a three-pronged approach. The previous management team had been "at war" with the union, Local 3 of the International Brotherhood of Electrical Workers. Nicholas took a far more conciliatory tone toward the union, improving performance and morale. The team also reached out and developed a much more positive working relationship with the city, its principal regulator. Manhattan Cable was able to get a rate increase—to $6 a month from $5—for the first time since the system went live in the late 1960s.[62] A direct-sales group created by the team helped drive revenues by going into apartment buildings, holding lobby parties, and developing stronger relationships with building managements throughout the city.

Home Box Office

One of the most valuable assets that Time Inc. acquired as part of the Sterling Manhattan transaction turned out to be its small Home Box Office business. But it took several years for Time Inc. management to conclude whether the unit was a lump of coal or a crown jewel. After all, carrying old movies and live sporting events on cable television was hardly a new concept. A major reason for Ross' initial interest in cable was to own outlets for the Warner Bros. vault of films.

HBO head Gerald Levin, left, worked with Andrew Inglis, president of RCA's satellite division, to lease transponder time on RCA's new Satcom satellite so that HBO could transmit its signal to systems operators around the country.

The specific idea for HBO came to Sterling Manhattan Cable's founder, Dolan, in 1971 while he was sailing on the *Queen Elizabeth II* to Europe on a family vacation. His initial name for the service was the Green Channel. Time-Life approved the idea in November 1971, seeing it as a vehicle for reusing some of its video programming and a much-needed shot in the arm.

In early 1972, Dolan hired the brilliant then-33-year-old Levin, an experienced negotiator with an impressive range of intellectual interests. Roughly a year after Time-Life approved the HBO concept, the network debuted on November 8, 1972.[63]

It was a rough maiden voyage. HBO began with its signal originating in Manhattan carried on a microwave feed to John Walson's Wilkes-Barre, Pennsylvania, system. A storm knocked out one of the microwave relays carrying the signal shortly before programming was to begin. A last-minute rooftop repair ensured that an unknown, but presumably small, number of subscribers could watch a New York Rangers hockey game from Madison Square Garden, followed by the movie *Sometimes a Great Notion*, starring Paul Newman. Two-step dance history was made the following spring when HBO's first original production, The Pennsylvania Polka Festival, was transmitted from Allentown on March 23, 1973.[64]

Levin was a good salesman, and by late 1973, the service was on 14 systems in New York and Pennsylvania. CVI owner Alan Gerry met Jerry Levin on one of Levin's tours through upstate New York touting HBO.[65] HBO received a significant boost during this period when Levin struck an agreement with Robert Miron, head of cable television operations for Newhouse Communications, for Newhouse's Eastern Microwave system to carry HBO in upstate New York. Levin's ties to Gerry and Miron forged during this period would play an important role in later multi-billion-dollar transactions between Time Warner Cable and its enterprises.

A weak U.S. economy did not help HBO's prospects for success. The 1973 Yom Kippur war and subsequent Arab oil embargo was already in the process of pushing the United States economy, and economies around much of the world, into the steepest economic downturn since the Great Depression of the 1930s. On top of that, the cable industry was still reeling, and its stock prices cratering, in the wake of an accounting scandal at industry leader TelePrompTer that surfaced in September—as if Kahn's legal troubles hadn't given the company a big enough black eye. The company restated its operating results, replaced existing management, and fired roughly 900 employees, which was more than the entire workforce of most cable operators.[66]

Experience it. Use your smartphone to watch a video of former Time Warner CEO Jerry Levin describing the development of Home Box Office.

Building Esprit de Corps: "We Were Peeling Potatoes over Here."

Emphasizing teamwork while recognizing individual talent has been a key to success for employees at all levels of Time Warner Cable. Nick Nicholas recalled the frantic pace and commitment by all involved in the turnaround of troubled Manhattan Cable in the early 1970s:

We became a team. We met all the time. Our offices were right next to each other. We had a lot of fun. We would go out to lunch and eat hamburgers every day and talk about the business. And we really invested in it.

We were all young, and in timing terms, the Marine Corps were over there fighting the battle of Iwo Jima and we were peeling potatoes over here. Nobody cared about us except [Time Inc. president] Jim Shepley. And he came down often. He had me come to his office once a week to talk about what was going on. And, of course, I brought Glenn [Britt] or I brought [someone else] depending on who was presenting. I would bring somebody with me. And so it was a chance for visibility for all of us junior bird people. And so it was a great experience.[67]

Scientific-Atlanta worked closely with cable systems operators to develop and install earth station antennas that were used as downlinks for satellite transmission of cable television signals.

Satellite Transmission

Industry visionaries had been promoting satellite transmission as the next big thing for cable television since at least the late 1960s. Rifkin's contractors from Bell Labs had warned in the late 1950s that it might pose a potential threat to cable by the mid-1960s. But even though there had been some modest successes in the form of special one-off satellite television broadcasts in the early 1970s, consensus seemed to be that the technology wasn't quite ready. And it was sure to be very expensive.

Levin, the newly anointed head of HBO, was willing to bet that the consensus was wrong. With the support of Time Inc. president Jim Shepley, Levin got the Time Inc. board of directors to vote in favor of pursuing satellite transmission of HBO in 1974. The hefty price tag—$7.5 million for a six-year contract for transponder time on an RCA satellite—[68] suggested Levin was an even better salesman inside of Time Inc. than he was out on the cable system circuit.

He had a willing accomplice in Sid Topol, president of Scientific-Atlanta. Topol was an industry leader in building and promoting giant dishes 10 meters in diameter to capture television signals bounced off of communications satellites and send them through cable systems anywhere in

Bob Rosencrans' cable system in Fort Pierce–Vero Beach, Florida, hosted Gerald Levin and other Time Inc. and HBO officials as they marked the HBO satellite transmission of the Thrilla in Manila fight between Muhammad Ali and Joe Frazier.

the country, or for that matter, the world. In partnership with TelePrompTer, he had actually demonstrated satellite transmission of a live boxing match at a cable industry event in 1973. Trouble was, the first-round knockout occurred before most industry viewers had even gathered to watch.[69] Topol convened a meeting in New York including himself, Levin, Hub Schlafly from TelePrompTer, Bob Rosencrans, head of UA-Columbia Cablevision, and Rifkin from ATC in the spring of 1975. The goal was to get systems to carry a live satellite feed of a major sporting event later that year. Even though Time Inc. had already taken an ownership stake in ATC, it was Rosencrans who took the lead in backing the satellite venture, designating his system in Fort Pierce–Vero Beach, Florida, as the test location. Levin announced the agreement with Rosencrans at the NCTA's annual convention in New Orleans on April 10, 1975. Rifkin followed suit shortly thereafter, choosing ATC's Jackson, Mississippi, system as the site for the Scientific-Atlanta downlink.

Thrilla in Manila

Talk about packing a punch. It's hard to imagine a higher-profile event to launch the first true commercial satellite transmission to a cable system than the heavyweight rematch between Muhammad Ali and Joe Frazier. Organizers chose Manila, Philippines, as the location of the event. But they timed it so that the opening round bell on the morning of October 1, 1975, would correspond with prime time on the East Coast of the United States on the evening of September 30. And HBO was the only service carrying the fight live in the United States.

Bob Rosencrans did the honors at the Vero Beach Holiday Inn, pulling the switch at 5:25 p.m. in front of more than 150 cable industry officials and local dignitaries to inaugurate the satellite era in cable television transmission. Monty Rifkin played a similar role in a downtown Jackson, Mississippi, hotel. Adding to the excitement at both sites was the fact that the fight turned out to be a heavyweight battle for the ages, with Ali triumphing over Frazier after 14 rounds. Two movies also shown that evening, *Brother of the Wind* and *Alice Doesn't Live Here Anymore*, didn't exactly match the pacing or drama of the boxing match, but they did showcase HBO's breadth of offerings.

Levin later admitted that he and Shepley had been challenged on the wisdom of using satellite transmission for cable programming so often from so many alleged experts leading up to the launch that he had practically worn a path in the carpet leading to Shepley's office to give him updates. He was beginning to have second thoughts himself. "You get enough naysayers that you begin to doubt yourself. So it wasn't until I saw the signal in Vero Beach that it was pretty clear that it was going to work."[70]

Cable television changed overnight. Systems lined up to install the giant satellite dishes. Even before the fight, TelePrompTer had placed an initial order for 50 Scientific-Atlanta dishes, getting a volume discounted price of $65,000 per dish. Time Inc. also extended millions in generous low-cost loans to smaller systems that might not have been able to afford the mammoth downlink dishes, in order to spur adoption of HBO.

As word of the service spread, customers across the country started demanding that their systems carry HBO. HBO's number of subscribers more than doubled to 275,000 by the end of the year.[71] They more than doubled again the following year. But a curious phenomenon soon became apparent: Customers were dropping the service almost as fast as new subscribers could be signed up.

TelePrompTer placed an early, sizeable order for Scientific-Atlanta satellite antenna earth stations that helped convince the antenna manufacturer that cable television was an industry worth investing time and money into meeting future equipment needs.

EVERLAST

1975

HBO's satellite broadcast of the famous Thrilla in Manila heavyweight bout between Muhammad Ali and Joe Frazier to a handful of U.S. cable systems on the evening of September 30, 1975, was a milestone in the evolution of cable television technology.

"If I Could Ever Get Into Anything to Do with Satellites, Do It."

Jeff King

An innate curiosity about the latest technology and a willingness to take risks had been key to the success of many in the rapidly evolving cable television industry. Jeff King was no exception.

After a tour of duty in the Army and then a stint in public access programming with ATC in Fayetteville, North Carolina, King had a chance to move back to his hometown of Terre Haute, Indiana, in the early 1970s and help revitalize the ATC system there. He and his wife were looking forward to putting down some family roots of their own, when a visit to the ATC system office in Battle Creek, Michigan, put King on a new career path. The manager in Battle Creek was a ham radio operator on the side, and regaled King with stories of bouncing signals off of the U.S.'s early Echo satellite. "He told me if I could ever get into anything to do with satellites, do it."[72]

It wasn't advice that King could put in practice in Terre Haute, until he received a call from Bill Brown, vice president of telecommunications systems in ATC's Denver headquarters, in 1974. "Would you be interested in going to Jackson, Mississippi, and adding a new signal called Home Box Office?" Brown asked King. King recalled thinking to himself, "Who the heck wants to move from your hometown to Jackson, Mississippi?" For all he knew about the South, Jackson might be located in a swamp.[73] But once Brown had mentioned that the signal was going to be carried by satellite, King said, "Yeah, I'm moving to Jackson, Mississippi."[74]

King and his crew arrived in Jackson in early 1975. The Thrilla in Manila was scheduled for September 30, and it was not a deadline that could be missed; Ali and Frazier weren't going to delay their rematch so ATC technicians in Jackson could debug their systems. King and his group worked closely with experts from satellite dish manufacturer Scientific-Atlanta—including a former NASA rocket scientist—to work out the specific telemetry required for the 10-meter dish. Manufacturing and site location snafus kept the crew working feverishly until 11 a.m. in Jackson on the day of the fight, but the signal was downlinked without a hitch, and the satellite era of cable television was born.

Monty Rifkin had told King that he would consider the satellite program a success if the Jackson franchise added 600 new subscribers at the end of a three-week free preview of HBO. In a clear indication of HBO's transformative impact, the preview period attracted nearly 3,000 new subscribers, five times more than Rifkin and been hoping for. That kind of immediate boost was repeated at ATC systems across the country, from Savannah, Georgia, to San Diego.

Focus on the Customer

Consumer research had not played a big part in most cable systems' business plans as of the mid-1970s. With the advent of nationwide programming, led by HBO, that was about to change.

Levin and the HBO team did an expert job in putting the fledgling service on the satellite and sparking nationwide interest among cable system owners. Their early bet on satellite technology was an industry game-changer, and put HBO on an industry-leading trajectory. Meantime, no one was paying much attention to the consumer receiving the HBO programs.

HBO had been on the satellite for little more than six months when Shepley contacted Nicholas. HBO was bleeding money, to the tune of millions of dollars a year, as a result of customer "churn." Levin remained chairman while Nicholas was sent in to revamp operations. He was given six months to demonstrate that HBO could turn a profit, or else it was likely that Time Inc. would shutter the business, satellite or no satellite.[75]

First, Nicholas overhauled HBO management and assembled the team that made HBO a household name across America within a few years. He brought in Austin Furst, the Time Inc. circulation whiz who was instrumental in the success of *People* magazine, as head of programming to negotiate with the movie studios. He also hired Michael Fuchs from the William Morris talent agency, who in turn recruited another talent *wunderkind*, Frank Biondi. Added on the operations side was Winston H. "Tony" Cox, a former *Life Magazine* executive, who later ran the rival Showtime/Movie Channel networks.

When the hit movie Jaws became available on HBO in the mid-1970s, system operator TCI saw its market penetration jump a full point, according to TCI head John Malone.

As the management team came together, they turned their attention to their customers, both the viewing public and major systems operators across the country. Contrary to the first couple of years of HBO history, the team began asking their consumers "What movies do you like? How would you like us to schedule? Do you like what we're showing you?" noted Nicholas. "Pretty soon we had a hell of a lot of intelligence about the consumer."[76]

"HBO and Cable TV Became Synonymous"

Nicholas also arranged for meetings with some leading cable operators on their home turfs, including Gene Schneider of LVO in Tulsa, Oklahoma, and John Malone and Bob Magness of TCI in Denver, to discuss what they recommended he do to improve HBO. Among their ideas were common-sense recommendations such as not starting one movie immediately after the next, and not scheduling a children's classic after a movie targeting mature audiences. Substantive changes in scheduling and promotion followed, and the public took notice. "All of a sudden the attitudes began to shift, fewer disconnects, and pretty soon we were actually net gaining subscribers," Nicholas said.[77] By the fall of 1977, HBO was showing a profit, and was spared the chopping block.

"We got our costs under control, really led by Austin; Michael started developing some rather unusual and interesting and very impactful non-film programming," Nicholas said. "I remember the first time I ever saw Robin Williams was on HBO, and the first time I ever saw Steve Martin was on HBO—brilliant, brilliant stuff that never could have been done on commercial television. Those are classic, just classic events."[78]

Cable system operations quickly realized that HBO delivered via satellite was their biggest moneymaker since the introduction of coaxial cable itself that had launched the industry. TCI's John Malone said that for several years, they weren't really selling cable; they were selling HBO.[79] When the 1970s blockbuster *Jaws* was released on HBO, Malone recalled, "The entire penetration of TCI went up a full percentage point ... People were signing up for cable to get *Jaws* ... HBO and cable TV became synonymous."[80]

"Truck Chasers"

Joe Collins witnessed the HBO impact first-hand—not just the impact on consumers, but the bottom line impact on the cable industry as well:

At HBO they had an expression for this, that as you first would bring HBO to a community you had people who were called truck chasers because they would literally see the cable truck go down the street. As part of your launch of HBO you'd have signs on the truck that would say HBO, and they would beat on the side of the truck to get the truck to stop so they could say, "When can I get this?" And so it was pretty phenomenal. It had a huge change in the industry.

The economics were fantastic because it did two things: not only did you all of a sudden have a new source of funds, you had a new $10 charge to go with what was now, let's say, a $7 monthly charge. You now got $17 from 35 percent of your customers, but you also all of a sudden had another big chunk of the market that had never bought from you before that bought both, so it had an unbelievable effect on the economics of the cable business.[81]

Quick to grasp the potential of satellite transmission of television signals, Ted Turner, shown here with Joe Collins, had his Atlanta television superstation transmitting via RCA's Satcom satellite by the end of 1976.

Programming Revolution

HBO pioneered a revolution in cable programming. Cable systems had been dabbling in locally originated programming since the 1950s, and wire services had been providing text feeds that systems had been displaying since the 1960s. But HBO was different. Virtually overnight, the industry had a source of programming that could be used nationwide for the cost of a satellite downlink and other fees. Others clamored to follow HBO's lead.

The latter part of the 1970s and early 1980s produced a flood of cable programming from a wide variety of sources. Some succeeded, others failed. Collectively, the burst of creativity not only changed cable television, it also marked the beginning of the end of broadcast networks' unchallenged control over television's destiny.

Superstations

Next in line behind HBO was Ted Turner, the most colorful satellite television superstar of the "superstation" era. Turner, building on experience in advertising and radio, owned two broadcast television stations in the Southeast as of the early 1970s, and had the right to broadcast Atlanta Braves games. (He bought the franchise outright in 1976.) As early as 1972, TelePrompTer systems in the region were carrying his stations' signals, and by the time of HBO's satellite debut, Turner's stations were reaching an estimated half-million cable subscribers.[82]

When a TelePrompTer executive alerted Turner in the spring of 1975 to HBO's plans to go on the satellite, he reportedly replied, "What's an HBO?" Quickly climbing the satellite learning curve, Turner recruited a young Western Union executive named Ed Taylor. Taylor, who'd led an earlier AT&T study of satellites, was tasked to run Southern Satellite Systems, a company Turner created to be the "common carrier" to distribute the satellite signals so Turner, a broadcaster, would not run afoul of existing FCC restrictions. The two worked almost nonstop for months to line up financing. On December 17, 1976, Turner's station, WTCG, began broadcasting via RCA's Satcom satellite to four cable systems.[83] The station's call letters were later changed to WTBS, for Turner Broadcasting System.

Turner, never one to diminish his own contribution to the industry's success, clearly saw the importance of satellite transmission to the future of cable television:

I said, 'Wait a minute, this is an antenna 22,000 miles up in space that can cover the whole North American continent and we can go point, multipoint, to every cable operator in the country, and if we put compelling enough programming on that satellite and give the industry something that people will be willing to pay for, then we can get cable operators to start in the major metropolitan areas, and we can wire the whole country and have a national medium, and eventually we'll get the 80 percent penetration, and then we compete with CBS, NBC, and ABC and make millions.'[84]

Cable News Network

Turner initially passed on the concept of creating a satellite cable news channel. Too expensive. Besides, he had built his broadcast empire in large part on programming, including sports and cartoons, that viewers saw as an alternative to news. But when he dropped the concept for a sports channel, only to watch with envy as the station that became ESPN went live in late 1978 following the launch of Madison Square Garden Sports in 1977, he realized he might suffer the same fate in cable news.[85]

Ted Turner's Cable News Network—CNN—launched on June 1, 1980, and changed the nature of global news coverage in just a few years.

The Copyright Act of 1976

ATC leadership played a key role in resolving one of the biggest regulatory, legal, and public relations thorns in the cable industry's side for more than 20 years: copyright. Bruce Lovett, chairman of the NCTA, made an impassioned plea to the group's 1975 convention that the time had finally come for the industry to settle the issue.

Lovett conceded that the cable television industry in 1974 had once again won the copyright battle in the U.S. Supreme Court, this time in TelePrompTer *v.* CBS. But he and other industry experts argued that after two decades of dispute, they had lost the war. They realized that if the industry was going to realize its full potential and be free of continued, nagging regulation, a copyright compromise was going to have to be reached.

And the venue for such a compromise was the U.S. Congress. Some smaller operators rebelled at the thought of compromise and seceded from the NCTA, but the major multiple systems operators (MSOs) held firm. In the fall of 1975 both houses of Congress passed the requisite legislation, and the Copyright Act of 1976 became a reality. The law took effect January 1, 1978.

A compulsory license was adopted for all broadcast signals, both local and imported signals from distant markets. Cable system operators would pay a royalty fee, beginning in July 1978, based on system size, market, and use of each imported signal. A Copyright Royalty Tribunal was created to adjust the fee schedule as required. Payments to broadcasters, sports interests, film studios, and producers were distributed by the Copyright Office of the Library of Congress.

A complex formula for payments led, predictably, to rounds of recriminations and lawsuits. But an important milestone had been reached. The Copyright Act raised the industry's image in the eyes of the public and also whetted the appetite of investors and lenders, who would be needed to fund the next phase of industry expansion.[86]

The NCTA provided a much-needed forum for the industry to hash out pending regulatory and legislative issues.

RIGHT
Ted Turner's willingness to bet his company on new ventures such as CNN rapidly made him a media and entertainment mogul to be reckoned with.

BOTTOM
Brian Lamb, founder of Cable-Satellite Public Affairs Network (C-SPAN), helped raise the cable television industry's public service profile when the channel began carrying live broadcasts of proceedings from the House of Representatives in 1979. Live coverage of the Senate was added in 1986.

In fact, both Time Inc.'s HBO and Warner Cable seriously weighed entering the cable news business in the late 1970s. Jerry Levin approached Reese Schonfeld, whom Turner hired to run CNN, in 1977 about HBO joining the Independent Television News Association (ITNA). Levin was interested in creating a news channel on HBO; ITNA, a company that pooled news coverage for small cable systems and was run by Schonfeld, seemed the perfect partner. As it turned out, ITNA's members didn't like the idea of aiding HBO, which they saw as a competitor. Time Inc.'s board also was cool toward what was sure to be another expensive venture.[87] Warner Cable CEO Gus Hauser and Katharine Graham of the *Washington Post* came close to an agreement to form a cable news channel in the late 1970s, but Graham, to her later regret, backed away. Cost, again, was the major concern.[88]

Ted Turner is famous for many things, but not cold feet. He was determined to make cable news work. So much so, he effectively bet his company on the concept. He found zero interest in the concept among cable systems operators at the December 1978 NCTA Western Cable Show in Anaheim, California. Undeterred, Turner and Schonfeld announced the formation of CNN at the May 1979 NCTA convention in Las Vegas. To finance the gamble, Turner sold one of his broadcast stations for $20 million. Slightly more than a year later, CNN went on the air on June 1, 1980. The venture was supported with profits from Turner's WTBS station for five years before finally posting a profit of its own in 1985. And by decade's end, it was setting the global standard for televised breaking news coverage.[89]

LEFT
C-SPAN expanded its commitment to public affairs programming when it added live coverage of national political conventions in 1980.

BOTTOM
Cable industry veteran Bob Rosencrans led his peers in providing financial support for C-SPAN and served as its founding chairman.

Cable-Satellite Public Affairs Network

Brian Lamb, Washington, D.C., bureau chief of *Cablevision* magazine, had interviewed many of the cable systems operators he stood to address at Washington, D.C.'s Mayflower hotel in August 1977. Now it was the turn of the operators—casting about for satellite programming ideas in the wake of HBO's success—to learn something from him.[90]

Lamb, who had previously worked in the White House Office of Telecommunications Policy, had been lobbying his publisher and others to support the idea of a cable public service channel focused on national affairs. Satellite transmission was tailor-made to suit the needs of such programming, Lamb argued. The fact that House of Representatives Speaker Thomas "Tip" O'Neil had earlier that year launched a test of live televising of House proceedings helped Lamb's audience envision what they might be able to offer subscribers.

In 1976, under the auspices of Cablevision, Lamb had distributed videotaped interviews of House and Senate leaders to a number of systems, ATC included, which had ponied up to support the effort. He also videotaped five days of House Commerce Committee hearings. The interviews and hearings were made available on HBO's satellite transponder during daytime hours when HBO was dark. (This arrangement may have played a role in Levin's interest in cable news programming.)[91]

By May 1978, Lamb had assembled a 22-person board of cable systems operators who chipped in to launch the Cable-Satellite Public Affairs Network (C-SPAN). The board was led by Bob Rosencrans, who wrote the first $25,000 check for the new venture. Dick Munro, head of Time Inc.'s video group, added his $25,000 check. ATC CEO Monty Rifkin sent Larry Howe, his head of cable programming, in his stead, but he did follow up with a $25,000 check as well.[92]

Lamb was operating on a shoestring budget, having raised just $450,000 to launch C-SPAN. Less than a year after its first board meeting, C-SPAN launched in March 1979, broadcasting live House of Representative sessions into 3.5 million homes served by 350 cable systems.

TOP
Warner Cable's two-way, interactive QUBE technology set a new technological standard for cable systems when it was introduced in December 1977.

BOTTOM
A group of boxing fans in Columbus, Ohio, used QUBE to call up a Roberto Duran fight.

QUBE provided customers with the ability to send responses back through their cable to a central facility using a handheld unit with 36 buttons, which was wired to a box that sat on top of the customer's television.

Warner Cable's QUBE

Network television devoted scant coverage in the 1970s to what was widely considered its country cousin, cable television. Any coverage at all tended to focus on the negative. But on December 1, 1977, the networks made an exception.

CBS's Walter Cronkite, the dean of network news anchors, looked into the camera and the homes of tens of millions of viewers and announced that a new cable service was starting. It promised to revolutionize the way Americans viewed, and used, television. It was the brainchild of Warner Communications, and it was called QUBE.

QUBE enabled true two-way communications over coaxial cable. Warner Cable addressed the problem of uplink interference or noise that had plagued ATC's system in Orlando a few years earlier by employing new computer-driven technology. Using a mammoth mainframe computer from Data General, the system scanned every QUBE subscriber's television hookup every six seconds. For an instant, what was known as a bridger gate was opened for uplink communication from a relatively small group of subscribers to the system. Orders for pay-per-view movies, for instance. Or voting on a local referendum or game show. Then it shut, all but eliminating the opportunity for signal noise to leak into the system. Each subscriber's connection was addressable, which enabled the company to know what every customer was watching, or ordering, in six-second intervals.

Such an intimate look at customers' habits was closely guarded. Unlike in the Internet era to come, no one would have dreamed of sharing such information with marketers or advertisers. Still, the "Big Brother" aspect did give some viewers pause, said Gerry Campbell.

"So many would come and say, 'I didn't watch that movie.' We had to watch how we did it, and eventually we had a change of policy—but when they called up, the service representative would say, 'Your television, right now, is on channel five, and you turned it off last night at 11:22.' Because we knew.[93] ... People used to put a towel over the TV in their bedroom because they thought we could watch them."[94]

The subscriber in the home held a controller with 36 buttons, one for each channel offered on the system. A precursor of today's remote control, it was tethered to a box attached to the television by a 15- or 20-foot cord. An enduring source of frustration for subscribers and technicians alike was the fact that customers were constantly tripping over the wire.

"I Want My MTV!"

On the programming side, Hauser assembled a team from the network and entertainment industries that developed some of cable television's greatest hits in the 1970s and into the early 1980s. They were driven by the need to develop content to fill the channels on the Ohio systems that weren't occupied by the networks, Public Broadcasting System, and a handful of independents.

One of Hauser's top priorities as QUBE was readied for launch was an all-day children's channel. He hired Dr. Vivian Horner, a professor who had been with the Children's Television Workshop, to execute the project. A few years later, Geraldine Laybourne was brought in to lead the effort. Originally called Pinwheel, the channel, renamed Nickelodeon, was ready for nationwide satellite distribution in 1979.

Brainstorming about content for preteens and teenagers, the team came up with the idea of a channel featuring music videos. Tapping Warner Music's talent pool to make videos was an obvious step. Examples of music videos playing on the BBC in the United Kingdom were another source of inspiration. MTV was born.

"The package of channels and services which eventually became the QUBE service foresaw virtually all the later developments of the cable industry," Hauser claimed, with only a slight amount of exaggeration. "For example, a culture channel was a precursor of the Arts & Entertainment Channel: a documentary channel was a precursor of Discovery Channel; a sports channel was a precursor of ESPN; a live, local channel was a precursor of local cable news channels; the QUBE adult channel was a precursor of Playboy and similar channels; the pay-per-view concept was developed as well as interactive features permitting viewers to select programs, vote, and purchase merchandise. The word *infomercial* was born to describe entire shows created by advertisers to feature their product."[95]

Warner Communications created cable programming history with the launch of its Music Television Network, MTV. Its inaugural cast of video jockeys, or VJs, pictured here, became a cultural touchstone and increased cable television's influence with teenagers and young adults.

Warner Cable raced to produce or acquire original programming to fill the channels of its QUBE system.

Hot Properties

By the late 1970s, cable systems were once again hot properties among investors and operators. The clear success of HBO, the technological advances represented by QUBE, and the steady stream of new programming available on cable systems made the industry a financial sector favorite. And advertisers were seeing the maturing industry, and the nationwide reach of satellite transmission, as a potential gold mine as well. For any operator who purchased a system in the late '70s and held on, it proved to be the purchase of a lifetime.

QUBE developed a strong subscriber base by 1979, and Hauser and Ross were ready to execute the second stage of their strategy: to use this cutting-edge technology to expand into major urban markets. To do so could cost hundreds of millions. Ross, concerned that the cost of expansion would place too great a burden on Warner Cable's balance sheet, turned to Rohatyn at Lazard Freres and said he needed to find a financial partner. One of Rohatyn's top candidates was American Express. QUBE whetted its appetite.

"Amex saw a lot of synergies with its merchandising and credit card operations," especially in the realm of in-home shopping over QUBE, "and they wanted to diversify," said Hauser. "They represented to Warner Communications an image that Warner Communications wanted to cultivate which would be helpful, not only financially, but in terms of being very, very white-shoe and clean. American Express was an icon," he added.[96]

American Express agreed to buy a 50 percent stake in the Warner Cable unit in December 1979 for $175 million. The company was renamed Warner Amex Cable Communications and a separate board of directors was created. Warner Amex initially lined up a $250 million line of credit to fund expansion efforts. That was later increased to $800 million, which at the time was the largest credit line ever granted in the cable industry. As an indication of the faith banks placed in Warner Amex, the credit lines were secured by the operating company, not its corporate parents.[97]

"We Had a Love Match"

With HBO in the black by 1977, Nick Nicholas made the case to Time Inc. management that they ought to own all of ATC. His strategy was to approach Rifkin and convey Time Inc.'s interest in purchasing more ATC stock, up to 20 percent, at which point they would have to include ATC's financial results with Time Inc.'s for accounting reasons.[98] Rifkin and his board had been enamored of Time Inc. since the initial investment in 1973, so they agreed. It all sounded perfectly reasonable. Time Inc. had been the friendliest of investors. "We had a love match going with Time Inc., and they with us," Rifkin later said.[99]

Nicholas called Rifkin several weeks later and asked that he join him, Dick Munro, and other Time Inc. officials in Denver at a taping of the *Phyllis Diller Comedy Hour* for HBO. Nicholas told Rifkin that Time Inc. had crossed the 20 percent threshold and now held 26 percent of ATC's stock. If it was going to consolidate ATC's earnings with its own, why not make the marriage formal and buy the entire company? Rifkin said, "Let me think about it."[100]

TOP
HBO Chairman Jerry Levin and President Nick Nicholas, center, in shirtsleeves, meeting with HBO regional and marketing officials in New York in 1978.

LEFT
With Warner Cable in search of added capital, and American Express anxious to tap QUBE's interactive potential to sell products and services, in 1979 the two formed the joint venture Warner Amex Cable Communications in 1979.

"If the Price Is Right ..."

The joke, sadly from his point of view, was on Rifkin. He had prided himself on not acting like some others in the industry and stacking his board of directors with cronies who would support him no matter what as he awarded himself huge amounts of stock and options. In fact, he was rather modestly compensated, having made a point of distributing cash and stock awards generously throughout company management. And he was only the fifth-largest shareholder.

From a founder's point of view, the timing of the proposed sale couldn't have been worse. The industry was clearly rebounding, and Rifkin had almost limitless faith in its future, as well as that of ATC's. Yet the price paid per subscriber in recent transactions hadn't budged much in a decade.

Rifkin's board was the problem. It was dominated by outside investors, led by Royal Little, who had given strong hints when Time Inc. made its initial investment that they viewed the top-drawer media company as a highly desirable ultimate owner. Daniels also would likely vote with those who wanted to cash out, and may have been one of the sellers of blocks of stock to Time Inc. as it boosted its stake.[101] Rifkin polled the board, which confirmed his fears. He then called Nicholas. "If the price is right we'll consider it," Rifkin said.[102]

Time Inc.'s first offer wasn't acceptable to the ATC board. But a second offer of $140 million was met with approval. Shareholders were offered the option of a tax-free exchange of Time Inc. shares for ATC shares, which Rifkin had advocated. The sales price came out to about $300 a subscriber, or little more than the price at which systems were changing hands in the 1950s and '60s. The transaction was announced in December 1978 and closed in November 1979.

Rifkin, sitting in his Denver office recently, not far from the site of ATC's first office building, held a note sent to him by Royal Little, dated June 25, 1987. In less than a decade the value of the ATC subscribers they had sold in 1978 had soared an astounding tenfold to an estimated $1.5 billion based on recent transactions in the business. Little acknowledged that Rifkin had tried to get him not to sell at the time. "This obviously was the worst mistake I ever made in business. Sorry I didn't listen to you! Sincerely yours, Roy."[103]

Focus on Franchising

The good news for ATC, which retained its name and senior management if not its independence, was that Time Inc.'s deep pockets came in handy very shortly. By the late 1970s, the industry was already in the process of gearing up for what at that time would be one of the largest, privately funded peacetime capital expenditure programs in history. With an array of program offerings, and enhanced systems technology, the industry had a compelling product to offer in urban centers and surrounding suburbs, which had excellent broadcast reception. Once these major markets were franchised, there weren't many domestic expansion opportunities left. The stakes were high, and no one wanted to be left out of the battle.

"One of the really big, major contributions that Monty and ATC made to the industry was bringing in a lot of sophisticated people ... They were able to join a lot of people who were also very entrepreneurial, and that made for a very good mix for a developing company and a new and growing business."

— Joe Collins, *Time Warner Cable CEO, 1989–2001*

By the late 1970s, Time Inc.'s video group was the most influential unit within the company and included the parent company's next generation of leaders. Video group executives, left to right: Gerald M. Levin, chairman of the board, Home Box Office (seated); N. J. Nicholas Jr., president, Home Box Office; E. Thayer Bigelow, president, Manhattan Cable Television; Bruce L. Paisner, president, Time-Life Films (seated); Thomas M. Girocco, vice president and general manager, WOTV; J. Richard Munro, group vice president.

Chapter 3

Growth and Innovation

The 1980s

In order to control costs that were snowballing out of control by the early to mid-1980s, Warner Amex Cable renegotiated terms of numerous franchises, including the system eventually installed in Milwaukee, pictured here, cutting QUBE from its product line and slashing the number of channels offered. Some systems were sold to competitors.

Time Inc. completed its takeover of ATC in 1979, but it was clear within a matter of months that the corporation's video group—with ATC as its rapidly accelerating economic engine—was taking over Time Inc.

The cable television industry was about to enter an unprecedented era of growth. ATC would also, in short order, account for an unprecedented amount of capital spending at Time Inc., running into the hundreds of millions of dollars, followed a few years later by an even greater contribution to cash flow. Warner Amex propelled growth at Warner Communications as well, even if its partnership with American Express unwound in the process by mid-decade. And by the end of the 1980s, the cable television businesses of both companies played a key role in their decision to jointly engage in another game-changing transaction.

Franchise Wars

Even before the sale to Time Inc. closed, ATC was leading the cable industry into the most competitive period in its history. In what became known as the franchise wars, the largest multiple systems operators (MSOs) competed for franchises in top urban markets across the country. In many markets, ATC and Warner Amex were locked in head-to-head competition from 1979 through 1981.

Industry pioneers had been forecasting just such a move for over a decade, though not without offering important caveats. Bill Daniels' advice at the 1968 NCTA convention to cable operators eyeing major urban markets? "Bring plenty of money."[1] With regulatory and legislative winds mostly at its back compared to a decade earlier, and with financing readily available based on the industry's potential, there was little holding the industry back. At least for those with solid financial backing and resources.

At Time Inc., Rifkin's top lieutenants—led by Trygve Myhren and Joe Collins—formed SWAT teams to lead the franchise effort into each major city. At the same time, they maintained their day jobs of running the existing operations to generate as much cash flow as possible. Back in Denver, Tom Binning, David O'Hayre, and their team worked feverishly, filling out spreadsheets by hand, to ensure that what was being promised during city council hearings around the country dovetailed with spending and revenue projections.[2]

Warner Amex Cable's innovative QUBE technology attracted engineers, like Harry Suri, left, from various fields and locations around the country. Steve Fry, right, rose through the ranks of the company's Ohio operations and became president of the northeast Ohio division.

Jim Gray led Warner Cable during a period of steady growth following the unwinding of American Express' investment in the cable company and through the early years of the Time Warner merger.

The same game plan was being executed at Warner Amex, noted Jim Gray, then-president of the company's national division and one of Gus Hauser's earliest hires. And everyone was playing for keeps. The economics of the industry forecast a future of front-runners and also-rans. As Gray noted, by the late 1970s, systems operators had concluded that "ultimately, this was going to be a business of size. There were scale economics to this that made it important that you be successful in franchising."[3] The last major untapped markets in urban America were up for grabs, and billions in revenue were at stake.

Educational and non-profit institutions saw the enhanced learning potential of QUBE's interactive technology. Kutztown State College in Pennsylvania used the two-way cable system in its Russian language studies classes.

The wars were stunning in the speed at which franchises were awarded in many of the markets in the country. The center cities and their many local suburban communities all sought franchise proposals at the same time. Warner Amex at one point had more than 100 applications pending.[4] Cities demanded futuristic services and the sophisticated infrastructure necessary to deliver them. In the view of some critics of the franchise process, this led to an array of pie-in-the sky promises and high-pressure marketing efforts by cable companies. These problems were not present in the Cincinnati competition in which he was involved, recalled Gray. "Cable technology was pushed to the edge of the envelope, but in the end, the Cincinnati cable system was built exactly as it was proposed."[5]

QUBE Appeal

Hauser and his team began developing the company's QUBE technology long before the first shot was fired in the franchise wars. Yet it proved to be the single-most potent competitive weapon deployed in the field. ATC's Collins recalled squaring off against Warner Amex and its crowd-pleasing two-way technology: "They began to use QUBE for franchising, and every time you would go up against Warner you went up against their QUBE stuff. And a lot of people would say to us, "What do you got? You know, they got this QUBE stuff; you got nothing.'"[6]

QUBE wasn't the only thing working in Warner Amex's favor, but it clearly had a deciding influence. "Again and again," *Cablevision* reported in December 1980, "the fact that Warner Amex is the company that has been using the only two-way interactive equipment up and running (excluding security systems) has lent credibility to bid claims that might otherwise be termed 'blue sky.'"[7]

By the end of 1980, Warner Amex already loomed large over the competition. It had landed major franchises in Cincinnati and virtually all of its surrounding suburbs, Dallas, Houston's northwest sector, Pittsburgh, and the suburbs of St. Louis. *Cablevision* identified Warner Amex and ATC as "by far the most active franchisers" in 1980,[8] but ATC was having trouble keeping up. During that same period, ATC added Kansas City; Erie, Pennsylvania; the City of Orange, California; Green Bay, Wisconsin; and Iowa City, Iowa.[9]

Although it lacked a functioning two-way system like QUBE at this juncture, ATC hardly had "nothing" to bring to the franchise wars. In addition to its 1980 "wins," it added Indianapolis in 1981, as well as franchises based in Honolulu and Austin, Texas, and continued to expand in smaller markets as well. By yearend 1981, in fact, ATC rose to the ranks of industry leader, with 1.8 million subscribers. (TelePrompTer and TCI were second and third with 1.5 million and 1.3 million subscribers, respectively.)[10]

Experience it. Use your smartphone to listen to a recording of former Warner Cable president Jim Gray describing the vital role played by Warner's two-way QUBE technology in winning cable franchises across the country.

"We Had a Lot of Firsts, Because We Had To"

ATC's top saleswoman in the early 1980s wasn't even on the company payroll. She did, however, control a significant amount of Time Inc. stock, and have influence over its board of directors.

Clare Booth Luce, widow of Time Inc. founder Henry Luce, was staying on the family estate in Hawaii in 1981 and introduced herself at a cocktail party to Carl Rossetti and Don Carroll,[11] chief financial officer and president, respectively, of the local cable system Oceanic Cablevision. She had heard that ATC had won the three-way bidding for the system, and enthused that the young men were "going to be part of the family!" Rossetti and Carroll were in the uncomfortable position of having to correct her. In fact, they were drafting a letter of intent with TCI, which had outbid ATC and Oak Industries, the third bidder.

The grande dame of American publishing wasn't the type to take no for an answer. A bit of behind-the-scenes lobbying on her part—and the fact that Hawaiian officials were chary of doing business with TCI in light of a well-publicized scandal in Jefferson City, Missouri, in which a company manager had threatened city officials[12]—led to a reopening of the bidding process. ATC prevailed in the second round.[13]

ATC's unconventional route to buying the Hawaii franchise, which today covers the entire state, was in keeping with the path-breaking practices that were, in fact, par for the course at Oceanic. "We had a lot of firsts, because we had to," said Rossetti, a current Time Warner Cable executive vice president and president of Time Warner Cable Ventures. "We were in survival mode all the time, and I don't think it's any different today. I mean, you have to be creative, you have to be innovative—in our industry and in our business—to basically keep things rolling."[14] In another first, Hawaii remains the only state with a single cable television operator.

Honolulu was "one of the first really urban markets built in the United States," Rossetti noted. "Very similar, really, at the same time, with Manhattan. So we had to be a little bit more creative" than those operating in smaller markets.[15] With the major networks having stations right in town, Honolulu residents had no problem receiving broadcast signals.

Oceanic's response was to introduce innovative programming. By the late 1970s, they were offering TBS and HBO via a 10-meter satellite dish located in a strip mall parking lot. They also had three movie channel options, dubbed Oceanic Theater. "We were the first company in the U.S. to offer multi-pay," Rossetti said.[16] Led by chief technology officer Jim Chiddix, who joined Oceanic in the mid-1970s after it purchased the small system he was working for on the western side of Oahu, Oceanic was also an industry leader in introducing addressable set-top boxes and automated billing, as well as creating a system for inserting local commercials into digital programming.[17]

Clare Booth Luce, right, wife of Time *magazine founder Henry Luce, left, was an effective, if unpaid, saleswoman for ATC in the 1980s as her behind-the-scenes efforts helped reopen the bidding process for the Hawaii cable franchise and led to ATC's bid being judged the winner.*

Seven technicians who joined the company at the same time in 1980 in Nebraska are still with Time Warner Cable. They include, from left, Rich Longwell, Larry Griffing, Dave Glantz, Don Bowens, Dave Bockman, and Dave Soldan. Not pictured is Bruce Colgrove.

Nebraska's Front Line

More than 30 years of service is an impressive achievement in its own right for any employee. But when seven technicians are hired for the same system in Nebraska at the same time and 31 years later they are all still with Time Warner Cable—there must have been something in the water!

With interest rates sky-high and gasoline prices close behind, the summer of 1980 was a tough time for most businesses. Fortunately for all concerned, the local cable system was hiring.

"When our group started, we were known as Cablevision and owned by Daniels and Associates. Then MetroVision, Advance/Newhouse, TWE-A/N, and now Time Warner Cable," said Rich Longwell, technical supervisor. As the name of the local provider has evolved, so has the cable service.

"We had 12 channels, and we were in on the first expansion to 23 channels ... how could we ever have enough programming to fill 23 channels?" he wondered at the time. Since then, the system has gone from 300 megahertz, to the current 870 mghz, supporting the triple play of video, Internet, and phone, with home security soon to be added to the mix.

Years ago, Longwell won a cable television slogan competition with the entry "more than just television." When asked how cable has changed over his career, his response was "cable is change." His slogan for today's cable service? "Cable, your gateway to technology."

A fresh-faced Glenn Britt transferred from Time Inc.'s Manhattan Cable division to ATC's Denver headquarters in 1980 to serve as chief financial officer of the cable company and help develop systems to better forecast the capital needs of ATC as it spent heavily to build out newly won urban and suburban cable franchises across the country.

"ATC Stood for All That Capital"

The money required to fund ATC's newly acquired systems certainly caught the attention of senior executives at Time Inc. back in New York. For a cash-rich publishing business whose customers paid subscriptions up front (and whose executives fretted over the number of typewriters they had to buy each year),[18] Time Inc. was shocked at the size of the projected capital expenditure figures coming out of ATC's Denver headquarters. The not-so-funny joke making the rounds of the Time-Life Building was that "ATC stood for All That Capital," Glenn Britt recalled.[19]

The numbers were adding up. In its unsuccessful bid for the Dallas cable franchise, ATC disclosed in bid documents in the spring of 1980 that it had capital commitments for "new builds" totaling \$121 million.[20] The capital budget for 1981 more than doubled to \$280 million and held steady at that level the following year.[21]

"The Youngest and Brightest Businesspeople"

Glenn Britt looked awfully young to be a CFO. At least that was Monty Rifkin's reaction in 1980 when Britt arrived in the ATC Denver office to replace Tom Binning, whom Rifkin had moved over to keep a tighter rein on ATC's construction division. Rifkin had sensed that the Time Inc. executives might be more comfortable with one of their own in Denver to help project spending schedules as anticipated franchising costs ballooned.[22]

Britt had to adjust to more than the altitude when he landed in Denver. "So Monty moved Tom into an operating job, and essentially said to the people at Time Inc., 'Who do you think ought to do this?' They sort of looked around, and I said, 'Well, I'll volunteer.' And I wasn't what Monty had in mind. And so here I am, I'm probably barely 30, and I'd worked at Manhattan Cable, but he was looking for a seasoned veteran, and I was a little wet behind the ears.

"I parachuted into Denver with people looking at me like, 'Who are you? What are you all about?' But I was able to make my way there eventually. And we were able to satisfy Time Inc., we could forecast capital, and all of that would work okay."[23]

ATC leadership learned that Britt and the other rising stars of Time Inc.'s video division weren't that far removed from the entrepreneurial spirit driving ATC. Munro had been given relatively free rein by Time Inc. CEO Richard Heiskell during the 1970s to build the video division as a much more risk-taking, free-wheeling business than the rest of the publishing empire. Britt recalled, "By the time we headed to the late '70s, that group clearly was the growth engine of the company. It was driving the stock price, and I think it did have all the youngest and brightest businesspeople."[24]

Munro and other video group executives aided ATC in the franchise wars, but it was a difficult process to manage. City councils and their advisors had been reading for years about the potential gold mine their markets would be for cable providers, and the amazing array of services cable could deliver. In addition, the terms for one franchise deal set the bar higher for the next city council. If one city system had 40 channels, the next city was expecting 52, and then more than 70, and ultimately systems with more than 100 channels. The number of channels, and promised local access studios and other services, kept rising. It was difficult to meet expectations that were changing by the month.

The competition to wire New York City beyond Manhattan was a case in point. ATC had done everything the city had asked, and then some, as it shepherded ATC's 40-channel franchise bid through the maze of approvals required for New York City's borough of Queens. But even after it was praised by a member of the City's Board of Estimate in March 1980 for sticking with the process, ATC stood by helplessly as its bid was rejected. In the time it took the bid to make its way through the city's tortuous process, rivals stepped in to promise more bells and whistles. Warner Amex promised an unprecedented 125-channel system and eventually won the franchise.[25]

"Rent-a-Citizen"

Most cities also had a preference for local involvement in cable franchise bids, though that was defined differently depending on the municipality. A process that became known as "Rent-a-Citizen" evolved, in which local investors, including representatives of minority groups, were included as investors in franchise bids, often on very preferential terms. Pittsburgh was a case in point.

ATC was the clear front-runner in Pittsburgh. In a nod to the needs of the city's poorer inner-city neighborhoods, ATC proposed providing 10 basic channels free of charge. It also arranged for a group of local minority investors to receive a 20 percent stake in the venture for a 5 percent capital contribution. But Warner Amex trumped that bid—giving away outright 20 percent of the venture to minority investors—and won the franchise, despite an in-person appeal from Munro.[26] The use of "Rent-a-Citizen" strategy was decried in the press, but there were no allegations of financial wrong-doing or fraud.[27]

Television personality and exercise guru Richard Simmons joined the ranks of entertainers singing QUBE's praises.

1977

AT&T's use of fiber-optic cable, being installed here in downtown Chicago in 1977 to carry trunk line phone calls, was an inspiration to Jim Chiddix, who used fiber-optic cable to carry television signals for ATC in Hawaii in 1985.

TOP
Terry O'Connell was one of several sales and marketing executives from Avon Products to join Warner Amex in the early 1980s to help bolster the cable company's sales and marketing effort. He rose through the organization and retired from Time Warner Cable in 2011 as executive vice president of Midwest operations.

BOTTOM
Kevin Leddy joined Warner Amex in 1980 from a staff position at the NCTA to help drive franchising efforts while also finishing business school. He switched his focus to marketing for several years, concentrating on supporting the company's technological innovations. He currently serves as Time Warner Cable's executive vice president for technology policy and product management.

"Marketing" Marketing

Steve Ross realized that he needed to beef up Warner Amex's marketing effort, which prior to the 1980s had been minimal at best. In 1982 he turned to Ed Carter, a senior executive at Avon Products, to create a true sales and marketing organization. Carter recruited about 35 executives out of Avon Products to come into Warner. Terry O'Connell was one of the recruits.[28]

O'Connell, Kevin Leddy, and others helped build the sales and marketing effort almost from the ground up. As vice presidents, O'Connell and Leddy reported directly to Lowell Hussey, chief marketing officer of Warner Amex. Together, working for Jim Gray, they traveled the country helping Warner Amex regions focus on marketing themselves and their products. There was nowhere to go but up.

"In the late '70s and early '80s, many of the system managers would take all of the marketing money that they had budgeted, not spend it, and drop it to the bottom line," said O'Connell, who retired from the company in 2011 as executive vice president in charge of the company's operations in the Midwest. "The only marketing that really took place was marketing that was done by the programming services. For example, HBO would market its product in each of our systems. That, combined with direct sales, was the extent of the marketing and sales done in most of these systems. We changed all that during the explosive growth years of the '80s."[29]

In addition to the firepower afforded by its QUBE technology, Warner Amex had one of the best field strategists in the industry in Dick Aurelio. The former deputy mayor of New York City under John Lindsay joined Warner Amex in 1979 as vice president of corporate affairs. He brought years of experience working with municipal bureaucracies to Warner Amex.[30]

Jim Gray huddled with Aurelio and Warner Amex CEO Hauser in Warner's offices in New York City in 1980.[31] As in Pittsburgh, Warner Amex appeared to be trailing ATC in the early rounds of the approval process for the cable franchise in Cincinnati and the 58 surrounding suburban communities. The three executives needed to come up with a new strategy, and fast.

Aurelio correctly sensed that the publicity surrounding "Rent-a-Citizen" required a new approach. Instead of giving ownership interests directly to individuals, as rivals proposed to do in Cincinnati, he decided that Warner Amex should bring in local institutions as co-investors. Free of charge, of course—and not just any local institutions. Aurelio lined up the Catholic Archdiocese, the YWCA, Xavier University, and the local public and parochial schools systems, among others.[32]

Once again, ATC had been outplayed, even though, as *Cincinnati Magazine* later reported, it had support from a local body formed to review the bids. On November 23, 1980, the Cincinnati city council approved Warner Amex's franchise bid.[33] In the surrounding suburbs, 53 of the 58 communities involved followed suit.[34] The Cincinnati institutions "eagerly split up 20 percent of Warner's local stock when Warner, through its benefactions, wrestled the cable franchise from Queen City Cablevision, a Time Inc. subsidiary that had been overwhelmingly recommended for the franchise by a specially appointed (and then largely ignored) cable advisory board."[35]

Bill Daniels, ATC, and other investors banded together to form Mile Hi Cable Television and bid for the Denver cable franchise. The Denver franchise, eventually awarded to the group in 1983, marked one of the last major skirmishes of the so-called franchise wars.

"Competition Is Feeding on Myth"

By early 1982, many in the cable industry had already sounded the retreat. Demands from cities were out of control, and the economics of franchising were getting increasingly hard to justify as well. The cost of wiring an urban home had jumped from an estimated $250–$300 in 1979 to $600–$800 by 1982.[36]

ATC, which built a record 10,000-miles-plus of cable plant in 1981 on its way to becoming the industry leader in number of subscribers, had called a halt to franchising that spring. An exception was in its own backyard, where ATC, Daniels and Associates, and other local partners had formed Mile Hi Cable Television to bid for the Denver cable franchise. The bid was bogged down by lawsuits and other issues, but a somewhat reduced version was eventually approved in 1983.[37]

Myhren cited Boston's 1981 demand for an interactive 70-plus-channel system, with estimated construction costs of $80 million to $100 million, as a turning point in the franchise wars. ATC walked away, as did many others. Warner Amex and Cablevision Systems Corp. were the only remaining bidders out of nine original applicants, with Cablevision eventually winning the franchise. Aurelio later told *Cablevision* magazine in 1982 that many Warner Amex executives had wanted to pull out of Boston as well.[38]

"The expectations of these cities are just unrealistic," added Myhren to *Cablevision*.[39] (Myhren had been named ATC president in January 1981.) The ultimate was the city of Sacramento, California, which demanded that cable bidders agree to plant 20,000 trees as part of their submission. "We are not going to endanger our shareholders or make false promises. Competition is feeding on myth," Myhren added.[40]

He also denied rumors making the rounds at the time that Time Inc. had put the kibosh on ATC pursuing further franchise deals. ATC was already looking ahead at building a war chest to combat possible competition from satellite television and the phone company. "It's not that we can't get money; it's should we?" he said. "We presented our revised five-year plan two months ago to keep embedded cable costs low because of foreseeable competition ... and because of the cost of money. They think it's a brilliant strategy."[41]

Changing of the Guard

Although Rifkin had a good working relationship with the Time Inc. executives to whom he reported, by 1982 he was ready for a change. Suffering from a case of "founderitis," he was tired of having to make serial presentations to different groups of Time Inc. officers and directors. He used to be called on by the Morgan Bank; now he was scheduling appointments with the Time Inc. CFO.[42]

In March 1982, Rifkin announced that he was resigning to start a venture to develop cable programming with Marvin Davis at Paramount. When that didn't work out, he formed his own firm, Rifkin and Associates, and once again went looking for cable systems to buy and upgrade. In 1983, he was named chairman of the NCTA. Trygve Myhren, the company's president and leading marketer, was named chairman and CEO of ATC, and Joe Collins, its operational expert, was named president.[43]

An Increasingly Civil Relationship

Harriet Novet joined Manhattan Cable in 1980 as the "franchise wars" that brought cable television to many of America's cities were still going strong. As often as not, she recalled, city leaders and system operators were at loggerheads, each trying to wring the best set of conditions out of the other, at times with scant attention paid to what was realistically achievable. As both sides have had more experience dealing with each other and working with the technology and services involved over the past three decades, and as several states have enacted statewide franchising, the franchise wars have been largely demilitarized to the status of negotiations, albeit with plenty of advocacy on both sides of the negotiating table.

A key factor in the increasingly civil relationship, says Novet, who retired in 2011 as regional vice president of public affairs in New York after helping structure a new nine-year franchise agreement with the city, is the rapidly evolving role of broadband communications in the life of the average American, and the fact that cable operators face competition from multiple sources in their major markets—unlike the cable landscape in the 1970s and 1980s.

Jack Gault, named president of ATC's Manhattan Cable system in 1979, played an instrumental role in molding the once-troubled Manhattan franchise into one of the most profitable divisions within ATC.

By the early 1980s, Hauser was also increasingly uncomfortable in his position as CEO and chairman for Warner Amex. American Express had purchased Shearson Loeb Rhodes in mid-1981 as part of its diversification strategy. Within months, Shearson and Amex executives were sniping at each other. Boasting of their leader Sandy Weill's prowess, Shearson executives denigrated Amex CEO Jim Robinson's decision-making skills, including his investments in cable television.

Hauser was feeling increasing pressure from the Shearson contingent on his own board to rein in Warner Amex. He had to push back on the board contingent and drive through the company's winning bid for the Brooklyn-Queens franchise in New York. Meanwhile, the costs of Warner Amex's rapid expansion were mounting, and QUBE wasn't proving viable as a standalone business. By early 1983, he had had enough, and like Rifkin, left to form his own firm, Hauser Communications, to buy and operate cable systems.[44]

Clustering and Decentralization

With franchising on the wane by 1982, ATC formed a cable investments group "to analyze and negotiate potential purchases, sales and trades of cable systems for improved financial and operational management."[45] Systems were grouped to form a critical mass in markets so that the subscriber base would become attractive targets for advertisers. The concept of what in short order was dubbed "clustering" was born, then adopted by other systems at roughly the same time as well. It continues to play a critical role in the way today's Time Warner Cable organizes its operations.

One of the first issues addressed by the ATC team that took over from Monty Rifkin—Myhren as chairman and CEO, Collins as president, and Britt as CFO—was how to reorganize an operation that had become much too centralized. They all felt that more decision making and understanding of how the business functioned had to be pushed down to the operating system level, not concentrated in Denver. There was also strong sentiment that systems should be grouped into divisions to enable shared services.

Two ATC systems, the recently purchased Oceanic franchise in Honolulu and Manhattan Cable, served as the poster children for decentralization at ATC.[46] Because it had been operating as a standalone business, Oceanic had all the functions in place, such as marketing, engineering, operations, and customer service, which would be required for a decentralized system within ATC. Manhattan Cable had also created what in effect was a division unto itself, with, for instance, its own engineering and marketing departments.

"We became a very decentralized operation. That freedom in the local operations resulted in a very, very strong uptick in business" companywide, said Jack Gault, who was named president of Manhattan Cable in 1979. In Manhattan alone, annual free cash flow went from roughly $1.5 million in 1980 to $45 million by decade's end.[47] Indeed, ATC was providing more than half of parent Time Inc.'s cash flow by the late 1980s.[48]

Changing Corporate Cultures

June Travis was put in charge of rolling the decentralization process out across the company, in what ended up being a five-year process.[49] The first division created in 1983 was Central Florida, consisting of former systems in Orlando, Kissimmee, Melbourne/Cocoa, and Ormond Beach. San Diego was identified shortly thereafter as the second division. Each was a minimum grouping with at least 100,000 subscribers.[50]

By the end of the year, 16 additional divisions had been identified, encompassing all of ATC's cable operations. Included was a national division, for "those cable systems not in proximity to other divisions."[51]

The decentralization process provided Britt with a valuable lesson in how well corporate cultures adapt to change. "I learned a lot from that, about how you take a culture that's very strong and change it to something radically different. That has stuck with me."[52]

Part of the lesson was that people can only take so much change. "I think that people in social institutions can deal with change, but only at a certain degree, and if the change gets to be too overwhelming, then I think things kind of blow up ... I don't know if I'm right or wrong, but I've kept that with me, including in my current job," Britt said in 2011.[53]

Warner Amex Revamps

Amex CEO Robinson, alarmed about the mounting costs of the cable business, helped recruit Drew Lewis, a former Secretary of Transportation under Ronald Reagan, to take Hauser's place as Warner Amex chairman and CEO. Lewis wasted no time in downsizing the company's efforts. He also moved company headquarters from the Warner Communications building at 75 Rockefeller Plaza in New York City to Blue Bell, Pennsylvania, outside Philadelphia, where he had a farm.

The company pulled the plug on QUBE as an ongoing enterprise in 1984. Costs were cited as the main reason for the decision. The programming had earlier been separated from the hardware operation and moved to New York. A number of systems continued to use QUBE, with the last being Cincinnati, where it remained in use until 1998.[54]

Industry's First Training Program

In 1981, ATC opened its National Training Center in Denver, the first major training center of its kind in the industry. The Center graduated more than 480 personnel in just its second year in operation.

The rapid expansion of its business put pressure on nearly every aspect of ATC's operations. To cope with the demand for more, and more qualified, personnel as the company expanded into additional and bigger markets with increasingly complex systems, ATC in 1981 opened its National Training Center, the first of its kind in the cable industry.

Employees from ATC's systems across the country could come to Denver and receive "the best technical and management training available in the industry," as the company said in 1983.[55] The National Training Center graduated 159 technical personnel—installers and service technicians—in 1981. That figure jumped to more than 480 in 1982.[56]

During this period, June Travis, working with finance professor Ron Rizzuto at the University of Denver, developed key aspects of a program that helped operating system general managers develop a better grasp of finance and balance sheet-related issues, which they might not have been exposed to in the field.[57]

By 1983, Trygve Myhren, who succeeded Monty Rifkin as ATC chairman and CEO that year, and his leadership team realized that the company had become too centralized and that more management decisions needed to be made at the operating system level. June Travis, right, took charge of administering the multi-year decentralization process.

The collapse of Warner Communications' Atari game business by 1983 added to the pressure on Chairman and CEO Steve Ross as he scrambled to raise the cash necessary to buy out American Express' interest in their struggling Warner Amex cable partnership.

Ponged

By 1984, Warner Amex CEO Drew Lewis had an added incentive to pare back on the cable business. Warner Amex's problems were exacerbated during this period by the stress put on its Warner Communications parent by a collapse in profits at its Atari game division. Atari, creator of Pong, one of the first home video games, had contributed about one-half of Warner Communications' profits in 1982. The following year demand evaporated just as Warner Communications had filled warehouses with machines it was not going to be able to sell. Atari cost its parent $533 million in 1983, and Warner Communications sold nearly all of its interest in Atari the following year.[58]

Lewis sat down matter-of-factly with city councils, which had grown used to making cable industry executives squirm during the franchise wars, and said no. No, they couldn't have the package that Warner Amex had negotiated in good faith. Upon review, it was simply impossible to build many of the promised systems at the promised price. Lewis, who had gone toe to toe with the Air Traffic Controllers Union during its landmark strike early in President Reagan's first term, was not one to be deterred by histrionics.

In Milwaukee, QUBE had been a major selling point for Warner Amex. Sorry, Lewis said, QUBE had to go. Too expensive. He then said that he would cut the number of planned channels in half and slash the access channels from 18 to six. Oh, and they would have to agree to a rate hike if Warner Amex was going to be able to build the system. Lewis began to renegotiate franchises with several other cities, too. When agreements could not be reached, he sold the systems. Most notably, he sold Pittsburgh, at the time the most profitable QUBE system, to TCI, and sold Dallas to Heritage.

TCI's Malone had been conspicuously absent from much of the franchise wars. He didn't have a parent corporation bankrolling him to the tune of hundreds of millions. And he consistently questioned the viability of QUBE and wannabe two-way offerings from other cable operators as largely untested and horrendously expensive. He preferred to stick with plain-vanilla cable offerings and run them to maximize cash flow. He had done just that in the suburbs surrounding Pittsburgh, so when Warner Amex put the system for the city back on the market, he was ready to step in and take it off Lewis' hands, basically at cost. Needless to say, nothing resembling QUBE was part of the TCI offering, which was reduced to 49 channels from 60.[59]

"Lewis' resolve convinced other cities that Warner Amex really had no alternative but to renegotiate or leave, and so they agreed to new terms," remembered Kevin Leddy, who at the time was director of marketing for the metro division of Warner Amex and today is executive vice president of technology policy and strategy for Time Warner Cable. "When Lewis sold Pittsburgh to Malone, it sent a strong message to Milwaukee and other cities that he was serious, and that they had to renegotiate."[60]

The Milwaukee City Council denounced the company and its executives, but they did not want to go the way of Pittsburgh. The odds of getting another operator to rebid the system at any point in the foreseeable future were about nil, and other cable companies would not buy Milwaukee given its onerous build requirements. Many operators were going through similar, if not quite as drastic, situations with cities around the country as the cost of building elaborate urban systems hit home.[61]

Drew Lewis, right, chairman and CEO of Warner Amex, and Jim Gray, center, president of Warner Amex's national division, received an award on behalf of the cable company's national division in 1984 from the Muscular Dystrophy Association for making the largest contribution of any cable television company in the country.

The Cable Communications Policy Act of 1984

By the late 1970s, the FCC had effectively removed most of the federal regulatory constraints on the cable industry. As the 1980s began, the cable industry's focus shifted to Capitol Hill, where it stepped up its lobbying for legislative relief from the largest remaining thorn in its side: local control of rates and franchise renewals.

The franchise wars had been a bruising experience. The industry-wide consensus was that something had to be done about the City Hall Caesars who had dictated such onerous terms to virtually every systems operator. Now was the time to strike back.

Reagan-era deregulatory winds were at the industry's back, and there was a great deal of optimism that something would be accomplished. This was so, despite the continued opposition of the powerful broadcasting networks and the regional Bell operating companies, which were being formed out of the process of breaking up AT&T, which took effect in 1984. At the same time, some leading cable systems operators refused to support proposed legislation, arguing that it didn't go far enough.

ATC and Warner Amex executives played pivotal roles in pushing the deregulation bill through. Myhren spent a good deal of time in Washington, D.C., in the early 1980s, both representing ATC and as vice chairman of the NCTA. He worked closely with Colorado congressman Tim Wirth, who was an important proponent of the effort as a member of the Telecommunications Subcommittee of the Energy and Commerce Committee drafting the legislation. At one point Wirth had the subcommittee's counsel and Myhren closet themselves in a conference room and craft some of the language that appeared in the final bill.[62]

Warner Amex's Lewis had been head of the Republican National Committee that helped propel Reagan into the White House. He helped line up support for the bill on the Republican side of the House.[63] Despite support from both parties, it wasn't until the final day of the 1984 session that The Cable Communications Policy Act of 1984 was passed. The bill, which placed a 5 percent cap on franchise fees, and provided for all but automatic franchise renewals, would take effect at yearend 1986.[64]

Unwinding a Partnership

By 1985, American Express had soured on its cable investment. QUBE, while chock-full of technological innovations, hadn't delivered as a vehicle for delivering financial services and other shopping experiences to the home. Amex wanted out, and they were looking for a buyer. A potential buyer was Drew Lewis and his management team, who wanted to do a leveraged buyout of Warner Amex. Management teams in many companies were doing LBOs in the mid-'80s, and Lewis, who saw the tremendous potential for cable in a post-regulation environment, saw the opportunity for his company.

Aware of the financial straits Warner Communications was operating under, Amex exercised a buy/sell clause that was part of its 1978 agreement to form Warner Amex. Under the terms of the agreement, one party could make an offer to buy out the other party. The other party was then required to sell, or buy out the instigator at the stated price. Amex's plan was to buy Warner's 50 percent and then sell the whole company to Lewis via the LBO, which they supported.

American Express offered to buy Warner Communications' one-half interest in the cable systems and programming for $450 million, widely considered to be a below-market price.

Warner Amex customer service representatives worked diligently during the 1980s even as they handled cascading customer complaints as the company frequently trimmed offerings to cut costs.

Amex was apparently gambling that Warner wouldn't be able to come up with the cash to match the offer, and therefore would have to sell to American Express. As Hauser recalled, "The market value of all the cable subscribers, all the not-yet-built or partially built cable systems and the programming business was valued by analysts as considerably higher than the American Express offer implied."[65]

Lewis was trying to raise $1.2 billion through Goldman Sachs, but he fell short and was not able to complete the buyout. Unbeknownst to Warner, American Express then cut deals with ATC and TCI to take the cable systems off its hands at the agreed-to price. The two actually formed a partnership to own the Warner Amex assets called, cleverly enough, ATCI.[66]

David O'Hayre, head of ATC's mergers and acquisitions strategy, led the team that negotiated the Group W deal in the mid-1980s.

However, Ross pulled together a deal on deadline to keep his cable systems in-house, though the cost was dear. In 1985 he raised the cash to meet Amex's $450 million price by selling the crown jewels of Warner Communications' cable programming, including MTV, Nickelodeon, VH1, and The Movie Channel, to Viacom for $690 million. Losing the programming assets was a blow, but the much-needed cash infusion above the $450 million price restored Warner's cable operations to a firm financial footing. [67]

Warner regained full control of the cable company, which became Warner Cable. Drew Lewis' team left for other industries. The headquarters moved to Columbus, Ohio, under the new president , Jim Gray, who had proved his ability to run financially successful cable operations throughout what was then called the national division.

Group W Deal Drives Clustering

ATC's leadership team put its clustering approach at the center of its acquisition strategy as of the mid-1980s. David O'Hayre, who had moved from controller to treasurer, was put in charge of the acquisition strategy. His team's mandate was to be opportunistic and build on existing ATC regional strengths when possible. "We wanted to take advantage of opportunities, whether with new franchising or acquisitions. We said, 'Here are our target areas, the places where we're already clustered, such as Charlotte or Manhattan; or Honolulu; Austin, Texas; so forth.'"[68]

Experience it. Use your smartphone to watch a video of former mergers and acquisitions strategy head David O'Hayre discuss the evolution of ATC's clustering strategy.

Cable Cassandras

Reprising the role of industry prophet he played in warning about the cost of building in urban markets, Bill Daniels cautioned industry leaders as the 1984 Act was about to kick in against getting too carried away with their newfound freedom. "Some operator is going to overplay his hand when it comes to deregulation and end up generating a lot of bad publicity and backlash for the entire industry. I don't know who it will be, but when you give people the freedom they've asked for, someone, somewhere, inevitably screws up."[69]

Trygve Myhren made a similar argument at a board meeting of the NCTA after the '84 Act became law. "I had a big argument in an NCTA Board Meeting with [John] Malone, and [Amos] Hostetter who got livid at me that I was arguing to restrain prices," he said.[70]

Myhren wouldn't be deterred by some who objected to his warning to other cable operators to act with restraint after deregulation. "It's not anti-competitive at all, it's consumer friendly. If you're talking about raising prices, it's anti-competitive ... I said, I feel strongly enough about it, I'm going to talk about it, because it's good for the consumer, it's good for our public policy position that we be restrained. I wasn't telling anybody what their price should be; I just said restrain the damn things. Because if you're not we're going to get reregulated."[71] By 1992, these cable industry Cassandras were proved right.

He didn't have long to wait. Westinghouse, via its Group W cable unit, bought TelePrompTer in 1980 for nearly $650 million and by 1985 had invested nearly $800 million more in cable television. That made it among the industry leaders with 2.1 million subscribers.[72] In 1985, the company decided to put its cable business up for sale. Westinghouse indicated that it didn't think the value of the business was reflected in the parent company's stock price.[73]

ATC, TCI, and Comcast led a consortium to buy the business, with each party willing to take about 500,000 subscribers each. The remaining half-million or so subscribers were split among companies led by cable industry veterans Bill Bresnan, Gus Hauser, Bill Daniels, and Leonard Tow. Experience and solid short-term financing led this group to prevail, even though another investor group made a higher offer. The winning offer was at less than $900 a subscriber. By the time the deal closed in June 1986 and the systems were apportioned to the investors, other cable operators were fetching closer to $1,400 per subscriber for their systems from buyers, underscoring the bargain price the negotiators for ATC, TCI, and Comcast had secured, O'Hayre recalled.[74]

Westinghouse's sale of its Group W cable business in 1986 to a consortium led by ATC, TCI, and Comcast hastened the industry trend already under way, in which ATC and others were clustering their systems into major market groups to achieve greater economies of scale.

Nick Nicholas, named president of Time Inc. in 1986, championed forming a partnership with Houston Power & Light called Paragon Communications to share the cost of buying and operating the systems purchased as part of the Group W transaction.

Paragon Partnership

ATC management was exuberant about the value they appeared to be creating with the Group W acquisition. They acquired TelePrompTer's former northern Manhattan system, adding it to ATC's existing Manhattan cluster based on the system in southern Manhattan. They also created new clusters in Tampa, Florida; El Paso, Texas; suburban Los Angeles; and upstate New York. Time Inc. executives, however, seemed to be getting nervous about the impact of the transaction.[75]

Nick Nicholas, head of the video group at the time, who was named president of Time Inc. in 1986, had a reputation for fiscal conservatism. He wanted to reduce the hit to Time Inc.'s net income as a result of the deal. So he sought an investor to share the load, particularly the acquisition of the northern Manhattan system.

So that the impact of its purchase of a portion of the Group W systems could be reported as an equity investment rather than having a direct impact on Time Inc.'s financial statements, ATC acquired the systems in equal partnership with a subsidiary of Houston Industries. The partnership, called Paragon Communications, held the northern Manhattan system as well as the other acquired Group W systems. Manhattan Cable officials and others argued, to no avail, that ATC and Time Inc. were giving away a significant portion of the upside potential of the deal by forming the partnership. Nearly a decade later, Time Warner acquired the partnership interest and other cable systems from HP&L, which had been renamed Houston Industries.[76]

The Group W deal put ATC firmly back in the industry lead in terms of number of subscribers—but not for long. TCI went on an acquisition binge, which put it back on top by the end of the decade. It added Heritage Communications, one-half of Storer Communications (with Comcast buying the other half), United Cable, United Artists, Taft Cable, and TKR Cable.[77]

Focusing on Fiber-Optics

Since at least the 1970s, futurists had been talking about applying fiber-optic communications—using strands of glass as thin as a human hair to very efficiently carry television signals on beams of laser-generated light—to cable TV. The technology was already widely in use to carry telephone calls long distances on AT&T's massive trunk lines. In July 1976, TelePrompTer set a CATV industry first when it used an experimental 800-foot long run of glass fibers to carry its HBO feed in northern Manhattan.

By the mid-1980s, ex-con Irving Kahn, TelePrompTer's irrepressible former CEO who had served his sentence in federal prison, invested in fiber-optics companies and became one of the technology's biggest proponents.[78] But fiber-optics was still a novelty item at mid-decade as far as cable television operators were concerned. That was certainly the consensus view of ATC's engineering department in Denver.

Fortunately for the company and the cable television industry, Jim Chiddix worked in Honolulu instead of Denver. If not, he might never have put the company in the vanguard of the biggest technological change to sweep cable television since the advent of satellite transmission of cable signals a decade earlier.

Chiddix, chief technology officer of the Oceanic Cablevision division in Honolulu, had been intrigued by fiber optics and their potential for transmitting cable television signals. The mountainous terrain of Oahu made getting signals from Honolulu to systems on the other side of the island especially difficult, even using a succession of microwave towers. In 1985, Chiddix hit upon the idea of running fiber-optic strands through a highway tunnel that the local telephone company was already using as a right-of-way for telephone lines.

The mountainous terrain of Hawaii's Oahu island makes line-of-sight transmission of television signals via microwave towers such as this one difficult. In 1985 Jim Chiddix, chief technology officer for the ATC franchise in Hawaii, used fiber-optic cable snaked through an existing highway tunnel to carry cable television signals across the island.

"This Was a Very Useful Tool for Cable Television"

As Chiddix was planning the fiber-optics run, he received a visit from Larry Jaynes, ATC's chief engineer. Chiddix and others credited Jaynes with curbing some of the company's more aggressive proposals during the franchise wars, saving ATC millions in the process. But he and his right-hand man, legendary engineer Austin S. "Shorty" Coryell, were conservative by nature, and told Oceanic president Don Carroll that they thought Chiddix should be using microwave transmission technology instead of fiber-optics.

Carroll, who had backed the trend-setting concepts his team developed out of necessity during the 1970s, wasn't about to change course just because Oceanic was now a division of ATC. As Chiddix recalled, "Don, to his credit, said, 'Look, you're my engineer; you figure out what the right answer is, and we'll just do it.' So we ran the fiber and just sort of ignored corporate, and the fiber worked very well. We got terrific signal quality across the mountains, and it really made an impression on me that this was a very useful tool for cable television" compared to microwave and other existing alternatives.[79]

TOP
Jim Chiddix, who moved from Hawaii to Denver in 1986 and was named ATC's chief technology officer, became the company's and the industry's leading proponent of game-changing fiber-optic technology, which ushered in the era of broadband communications.

RIGHT
In the mid-1980s, Louis Williamson helped engineer revolutionary breakthroughs in the use of fiber-optics. Today, he is Time Warner Cable's senior fellow of engineering and is one of the industry's leading experts on communications technology convergence.

"Aha" Moment

In 1986, Chiddix agreed to move to Denver from Honolulu to replace Jaynes, who had retired. One of Chiddix's first jobs was to close down a number of existing research and development efforts and pull together a research team to focus primarily on fiber-optics. He recruited the company's chief engineer in Kansas City, David Pangrac, who had also been working on fiber-optic transmission, to join him in Denver. One of the talented existing engineers in Denver that he redirected toward fiber-optics research was Louis Williamson. A consultant named Herzel Laor, a physicist researching optics and electro-optics, also played an important role on the team.

The Denver engineering team saw fiber as the answer to a twin set of challenges facing the cable industry. On the one hand, coaxial cable, with its string of amplifiers boosting signals every 1,000 feet or so at best, was reaching the signal-carrying capacity inherent in the technology. Significant increases in cable's ability to sustain its capacity over long distances were needed if the promise of an ever-expanding array of channels was to become a reality. In addition, ATC and other cable companies were bracing for competition from satellite television companies within a few years. They broadcasted digital signals directly into subscribers' homes.[80]

Experience it. Use your smartphone to watch a video of former CTO Jim Chiddix describing the process leading to his appreciation of how the use of fiber-optics could revolutionize the cable industry.

Beginning in the 1980s, as more cable systems, especially those in suburban locations, required cable to be buried, installers and technicians encountered a host of new challenges as they navigated around water and electrical utility lines as well as sprawling tree-root systems.

"The problem was that lasers weren't good enough yet to take the whole radio spectrum from the coaxial system and ship it through a fiber and then reconstitute it with acceptable noise levels," Chiddix said. "So we began talking to laser manufacturers about better lasers and the limits of laser noise. And we found some people who were doing some cutting-edge things with lasers, in military applications and telecommunications applications."[81]

Focusing exclusively on fiber-optics, the team made amazing progress. "I arrived in late '86. By mid '87, Louis [Williamson] on the bench, in the lab, had a fiber link up and going that was running, I think, 40 channels through maybe 10 kilometers of fiber."[82]

That was the "aha" moment for the team, Chiddix recalled. "Because if we could do that a little better with some more channels and a cheaper laser, we would have something that we could really put in the field and begin to change the architecture, get a lot more channels, a lot better reliability, and have something that could take us beyond where satellite could go. You know, this competitive satellite that was on the horizon."[83]

No "Secret Weapon"

"We thought about keeping it proprietary and having it be ATC's secret weapon," Chiddix noted. "But ATC really didn't have the critical mass to make this ... cheap. We really needed the whole industry to embrace it, we thought. So we talked to other cable operators about it. And we published papers that were as broad as we could make them, not only to educate, but to try to make it difficult for people to get blocking patents—to set up roadblocks or tollbooths to make it difficult to cost-produce this technology."[84]

1986 IPO Values ATC at $1.6 Billion

Reasoning that investors were not adequately making a distinction between parent Time Inc.'s focus on increasing annual earnings per share and ATC's strategy of borrowing heavily to build cable systems, Time Inc. officials thought it made sense to sell a portion of the cable company to the public. That would lead investors and Wall Street analysts to value ATC on its merits as a capital-intensive business and to highlight the difference between the subsidiary and the parent Time Inc.

ATC went public for a second time in 1986. It was one of the largest equity offerings of the era. Time Inc. sold a 20 percent stake in ATC for over $300 million, valuing the cable company as a whole at $1.6 billion, or just shy of 10 times what Time Inc. paid to take ownership of ATC in 1978.

While ATC executives may have anticipated using ATC stock to finance acquisitions in the cable industry, because Time Inc. would have lost its ability to consolidate ATC for tax purposes if it owned less than 80 percent of the company, using its stock for those purposes was unlikely.[85]

The good news was that as the '84 Act deregulating key aspects of the cable industry took effect at yearend 1986, the investment prospects brightened considerably for the industry and ATC. Like most leading cable systems operations, ATC had little trouble borrowing from banks to fund further acquisitions.

The team quickly coined the term *hybrid fiber coax* (HFC) to define the new technology. Creating a totally new system using just fiber-optics would be prohibitively expensive. "The idea was that there was all this coaxial cable already in place, going to all these homes. And we wanted to reuse that, because replacing all of that would be staggeringly expensive," Chiddix said. "But we wanted to get within maybe a kilometer of every home with fiber and have clusters of about 500 homes fed with coax. And those very short coaxial systems would work very well and support a lot of channels, be very reliable, and then have a passive fiber connection back to the headend."[86]

The ATC engineers quickly saw the potential for broadband applications in fiber-optics, Chiddix said:

Dave Pangrac and I gave a lot of speeches and did a lot of proselytizing around this, and as we did it, it began really occurring to us that this did more than reduce amplifier cascades. It also segmented the cable system into much more manageable pieces. If you were ever going to do things like telephone service over cable or data service over cable, or the ultimate dream, deliver an individual video channel to an individual home, this began to open that door. It began to change the architecture or the topology of cable in a way that could open that door. That really began to pick up steam.[87]

"We Were Like the Google of Our Era"

The leadership of the Time Inc. video group, including ATC, HBO and Time-Life Films, held an off-site retreat in 1983. McKinsey & Co. had prepared a study of the group for the occasion. Over the previous five years the group had grown from next to nothing to about $500 million.[88] McKinsey projected that the group's revenues were likely to double over the next five years to $1 billion. So far, so good.

Then the lead McKinsey partner heading the presentation delivered the bad news: Their research couldn't turn up a single example of a company that had a similar growth rate without suffering a serious setback at some point. Glenn Britt recalled the ominous warning: "Something bad is going to happen that you guys aren't anticipating. And we don't know what that is, but something's going to happen."

The assembled video group leaders scoffed at the idea. "We were like the Google of our era," Britt said. "It was like, 'We're growing like a rocket ship and we're all smart and you don't know what you're talking about.'"[89]

VCR vs. HBO

That "something bad" turned out to be the video cassette recorder, or VCR. VCRs had been introduced in the 1950s, but the professional-grade machines were refrigerator-sized and cost tens of thousands of dollars. Even when versions for the home market were introduced in the 1970s, the prices still topped $1,000 a unit. HBO initially hadn't taken them seriously as a competitive threat, even though they continued to get cheaper and more compact by the year.

By the mid-1980s, the threat posed by VCRs couldn't be ignored. Once the Supreme Court ruled in January 1984 (in what was known as the Betamax case), effectively opening the door to home taping of a movie using a VCR for personal use, consumer demand spiked. By 1987, nearly one-half of American homes had a VCR. Movie companies kept the price of movies on videotape high to encourage in-theater viewing. A major result of the pricing strategy, however, was to spark a boom in video rental outlets across the country.[90] Many worried that the more consumers rented movies to watch on their VCRs, the less they would tend to subscribe to HBO and other pay services.

Early video cassette recorders (VCRs) were cumbersome and expensive, but the new technology became a serious threat to cable programmers during the 1980s as units became more compact and affordable, and court rulings removed many of the legal uncertainties concerning the personal recording of video programming.

The potential impact of competition from VCRs was serious in its own right. It was made worse by a series of management decisions at Home Box Office that saddled the unit with high fixed cost contracts with movie studios, rather than tying contracts to the number of HBO subscribers. Infighting among the group of HBO executives, who reported to Jerry Levin at the time, made a bad situation worse.

Nick Nicholas was CFO of Time Inc. in 1984 when this issue came to a head. Time Inc. CEO Dick Munro asked him to serve as head of the video group and had Levin move into the role of Time Inc. chief strategist. Nicholas tapped two younger executives to serve as his assistants—Glenn Britt and Jeff Bewkes.

Nicholas decided to promote Michael Fuchs, HBO's number two, to CEO of the business, replacing Frank Biondi. But Nicholas also realized that HBO would benefit from senior management that really understood the cable business. After receiving a strong recommendation from Britt and others, Nicholas named Joe Collins, president of ATC, as president of Home Box Office.[91] Britt was named CFO of Home Box Office.

Collins ran into Biondi in front of the Time-Life building on Sixth Avenue the day he was reporting to work. "He came up to me and he said, 'Congratulations.' I said, 'Thank you'. And he said, 'Think of yourself as the mayor of Paris in May 1940.'"[92]

Despite the graveyard humor, Collins and Britt were able to renegotiate contracts and cut costs at the division. It was one of the first significant rounds of cutbacks at a Time Inc. business, where such reductions were rare, and left some raw emotions in its wake. "I was known as 'Glenn Brute' for a while," Britt said, "but we got HBO on good footing, and it really has done very well ever since then, but it was a dramatic event at the time."[93]

In addition, despite the early concerns about the impact of VCRs on pay services, the greatest growth in pay services came following the VCR's introduction. And the functions offered by VCRs were in key ways an early attempt to meet consumers' demand for control over getting access to content to suit their schedules, not those of cable networks.

ATC to Stamford, Connecticut

When Nicholas was named president of Time Inc. in 1986, he was considered Dick Munro's heir apparent as CEO. Nicholas had been very supportive of the decentralization push at ATC. But he also felt that ATC's headquarters needed to be moved closer to Time Inc. in New York.

By 1988, Time Inc. had owned ATC for a decade. Despite the public offering of just shy of 20 percent of ATC's stock in 1986, Time Inc.'s controlling interest led it to continue to consolidate the cable unit's results with those of the parent company. However, the rapidly growing cable company, which was accounting for more than one-half of Time Inc.'s cash flow by 1988, was still being run as if it were a separate company.

Trygve Myhren and Nicholas were both strong-willed executives. They had butted heads on a number of occasions concerning issues such as Myhren's desire for the company to take a more integrated, aggressive approach to cross-selling advertisers among its print and cable properties. Nicholas didn't pursue that opportunity, and Myhren felt that his team's input hadn't been considered in an ill-fated attempt to produce a cable viewing guide that ended up costing Time Inc. tens of millions of dollars.[94]

Their differences came to a head in 1988 when Nicholas announced that he was moving ATC's headquarters from Denver to Stamford, Connecticut. It was close to New York while still offering much cheaper space for ATC's offices than the city. Myhren parted ways with ATC and stayed in Denver. To run ATC going forward, Nicholas brought Collins back from Home Box Office and made him CEO, and Jimmy Doolittle was named president. Ultimately, only about 10 percent of the 600 headquarters staff in Denver relocated to Stamford.[95]

Time Inc. Looks for a Partner

Jerry Levin lost out in the race to succeed Munro as Time Inc. CEO in 1986 when Nicholas was named president. But he embraced the role of strategy chief, even if there was no clear blueprint as to just what that role entailed. He quickly determined that the position would give him a platform to advocate a "transforming transaction" that could propel Time Inc. to the front ranks of mainstream entertainment, building on its strengths in cable television at ATC and HBO's programming.[96]

Levin, like many on Wall Street and in the investing community, had concluded that Time Inc. was stuck in its conservative, predictable ways and was hopelessly inward looking. Even Dick Munro had sounded out other publishing empires, including Dow Jones, Knight-Ridder, and Gannett about possible combinations in 1985. CBS, also fearing a takeover, discussed a possible merger with Time Inc. as well, but no one could resolve the fact that the combined company would have to sell either its broadcast or cable properties. Time Inc. officials also held merger discussions with ABC/Cap Cities and Paramount Communications.[97]

With inflation firmly under control and the Reagan-era deregulation environment carrying the day in business and in the courts, Wall Street was on a roll by the mid-1980s. Low interest rates created a period of cheap money that made it easy for companies and corporate raiders to borrow heavily to finance acquisitions. Drexel Burnham Lambert's junk-bond business was providing ready cash for companies that might have been frozen out of the debt markets in more cautious eras.

The question hanging over Time Inc.'s executive offices was whether the company would be able to find a compatible partner, or would Time Inc. fall prey to the next debt-fueled deal on Wall Street? Wall Street's back-of-the-envelope estimates put Time Inc.'s break-up value, or the value of all its businesses if they traded separately, at twice what its shares were trading for in 1986, and maybe more.[98]

The executive team of Munro, Nicholas, Levin, Britt (who became treasurer of Time Inc. in late 1986, and vice president of finance in early 1988, replacing Thayer Bigelow, who became president of Home Box Office and later head of cable investments at Time Warner Cable), and Phil Lochner, general counsel, felt as if they were under siege. "My recollection was that [at least] once a week, we would hear a rumor that some raider was coming after us ... so we were busy trying to not be taken over, at least not for too low a price," Britt said.[99]

Warner in Focus

Warner Communications inevitably appeared on Time Inc.'s dance card of potential suitors. The Time Inc. team had considered pursuing a strategy of buying a movie studio to give it more content to distribute, which made Warner Bros. an intriguing possibility. Finding some form of combination for their largely complementary cable systems to gain even further economies of scale in cable also was an attractive proposition.

Nicholas and Ross met initially in early 1987 to discuss how the two companies might work together, or fit together. The discussions were preliminary, and talks with other potential suitors continued as well.

After years of on-again, off-again talks, the merger of Time Inc. and Warner Communications, announced in 1989, brought together Warner Communications chairman and CEO Steven Ross, rear, with Time Inc. president Nick Nicholas, left, and chairman Dick Munro, in an uneasy leadership alliance presaging years of management turmoil at the helm of the combined entity.

By late summer, however, Nicholas had become convinced that Warner was the one. Flying back from a family vacation in Kenya that August, Nicholas couldn't sleep during an overnight stop in Zurich. He went down to the hotel bar and started writing down on sheets of paper from a hotel notepad all the advantages of combining Time Inc. and Warner. He gave the notes to Levin, whom he acknowledged as the better communicator, once back in New York. Levin, adding his touches, produced a memo for Munro that outlined the Time Inc.–Warner deal.[100]

For the balance of 1987 and into 1988, the companies considered a deal short of an outright merger. They weighed a complex corporate combination that would leave both Time Inc. and Warner publicly owned companies, with publishing consolidated into Time Inc.; music assets into Warner; and the Warner Bros. studio, HBO, and the two companies' cable businesses combined and owned 50-50 by Time Inc. and Warner. That approach was dropped finally when the companies realized it would have triggered major tax bills for both parties.[101]

Executives from business units of Time Inc. and Warner Communications interacted frequently at joint industry events prior to the merger of the two companies, including this 1988 Movietime press conference. At left are Ed Hamowy of Warner Communications and Jeff Bewkes, CFO of Home Box Office, while Warner Cable's Paul Jones is on the far right. Second from right is Larry Namer, formerly of Manhattan Cable TV's Commercial Services department, who created "Movietime," which later became "E!"

Making the Merger Case

A confidential strategy report Time Inc. management had commissioned was distributed internally in June 1988. It stated, "The company is perceived on both Wall Street and internally as uncreative, overly cautious, investor-driven and risk averse." The report added that "senior management has not communicated a galvanizing vision of the company's future growth opportunities." It also said Time Inc. should increase its "ownership of video rights and creation of video programming."[102]

That month, Nicholas met with Ross and said that Time Inc. management felt it was time for the two parties to pursue an outright merger.[103] Ross, who had been pushing for a merger for months, was enthusiastic. With the strategy report supporting their argument that Time Inc. had to make a transformative transaction, Nicholas and Levin, with Munro's blessing, began making the case internally at Time Inc. for a Warner merger.

The business case for the merger was the easiest part of the selling job internally at Time Inc. Real synergies in combining the cable assets of each company, and hoped-for synergies in the entertainment and publishing businesses, supported bringing the two companies together. Cultural and leadership issues were a higher hurdle.

Larger-than-life, brilliant, and free-spending bon vivant Steve Ross had built Warner Communications from practically nothing to a company with a market value of $5.6 billion in 1988. He was Warner, and he had a great deal of influence with his board of directors. That made Time Inc.'s independent-minded directors nervous. Ross had been implicated but never indicted in a scandal surrounding Westchester Premier Theater, in which some former associates and employees were convicted of racketeering and fraud.[104] And Munro and his team learned during this period from a secret "mole" on the Warner board that Ross was battling prostate cancer, which eventually became public knowledge.[105]

"It's a Good Feeling That the Customer Can Learn Something from My Advice."

Cesar Torres

Cesar Torres worked at three other cable companies—the last was Adelphia—before coming to the company in 1983 as an installer in Santa Monica, California. It was a good fit, says Torres, now a headend technician II.

Things are changing so fast ... when I started it was only one-way service (video). Now we have two-way services. We can provide more services to our customers; they have more choice in the programs they can subscribe to. When I was working as an installer at Hermosa Beach in 2002, an elderly customer had an on-the-ground cable issue.

It was after hours, and our construction department was closed, so I actually changed the cable for him. He was like, 'Wow!' He was so thrilled he said he'd do anything for us for making sure he had TV running in his home without any break!

What's great about TWC? The company cares for its employees and takes care of its customers. The company is always there for us. It's involved in the community—that's another thing I like about the company. I don't directly deal with customers anymore because I'm a technician, but whenever I get a chance, I like to help out. I answer any technical questions they have. It's a good feeling that the customer can learn something from my advice.

The merger nearly foundered on the issue of who would run the combined company. Munro was adamant that Nicholas become the CEO of any merged company. This requirement reportedly played a role in quashing the interest of certain merger partners. Ross initially balked at the idea. He eventually agreed, in the form of a legally nonbinding understanding, that he and Munro would be co-CEOs until Munro's previously announced retirement in 1990. Nicholas would then step into Munro's shoes as co-CEO and become sole CEO at yearend 1994, though Ross would remain as chairman.

The Time Inc.–Warner merger was announced in March 1989 as an exchange of stock valued at $18 billion. On June 7, two weeks before the deal was to be voted on by shareholders of Time Inc. and Warner, Paramount Communications Inc. made an unsolicited $10.7 billion cash offer for Time Inc., which was later raised to $12.2 billion, or $200 a share—a significant premium to the company's share price prior to the Paramount bid. Time Inc.'s executive offices were converted to a war room where executives operated in lockdown mode. Assistants were not only sent out for food, they were also sent to buy shirts and underwear as days turned into weeks.[106]

Time Inc. rejected the Paramount offer, arguing that its board didn't have to consider the bid since the merger with Warner was a combination of the two companies. Time Inc. was not putting itself up for sale. Time Inc. and Warner then restructured their combination as a purchase of Warner by Time Inc. for $14 billion. That left newly minted CFO Britt scrambling to line up financing for the deal, and doing so in an unprecedented three weeks.[107] Time Inc.'s decision was upheld by the Delaware Chancery Court, and seen as supporting a corporate board's authority to act in the long-term interests of a company. The case is still cited as a major corporate law case.

Time Inc.'s litigation team was headed by Robert Joffe of Cravath, Swaine & Moore in New York, the long-term lead outside litigation and corporate counsel for Time Inc. and later Time Warner and Time Warner Cable. Marc Lawrence-Apfelbaum, a former Cravath associate who worked with Joffe on many matters for both companies, joined ATC during the Paramount battle to help lead ATC's litigation efforts during the case. He was later named Time Warner Cable executive vice president, general counsel, and secretary.

Connecting Cable

Combining the leadership and operating components of Time Inc. and Warner Communications into a smoothly functioning whole, while working off the mountain of debt used to finance the deal, consumed a great deal of the newly merged company's attention and focus for a number of years. In addition, the highly successful cable division, along with the rest of the industry, would be whipsawed by successive waves of re-regulation and deregulation to a degree unparalleled in modern business history. And even as those trends were playing out, Time Warner Cable in the 1990s played a leading role in ushering in the modern era of digital broadband communications.

The combination of Time Inc. and Warner Communications was heralded as creating a leader in global communications, media, and entertainment for the 1990s and beyond.

Time Inc. and Warner to Merge, Creating Largest Media Company

Print and Movie Giants Are Combining to Fight Foreign Competition

By FLOYD NORRIS

Time Inc. and Warner Communications Inc. announced yesterday that they plan to merge, creating the largest media and entertainment conglomerate in the world.

Time's chairman, J. Richard Munro, said the new company would seek to grow even larger by acquiring other businesses.

Time is a leading book and magazine publisher with extensive cable television holdings, and Warner is a major producer of movies and records and has a large cable-television operation. The merger would create a new company, Time Warner Inc., with a stock market value of $15.2 billion and revenue of $10 billion a year.

A Place in 1990's

The merger would insure Time Warner a place in the 1990's as one of a handful of global media giants able to produce and distribute information in virtually any medium. The companies

mann A. G., a privately held German publisher known primarily for its book division, as the world's largest communications company in terms of revenue. Bertelsmann's 1987 revenue was more than $6 billion.

The merger was unanimously approved by the Time board, but there was one abstention on the Warner board. A Warner official said Herbert J. Siegel, chairman of Chris-Craft Industries, who is a frequent critic of Mr. Ross, was the holdout. Mr. Siegel could not be reached for comment.

The agreement is subject to approval by shareholders and by Government

The Massa family of Whitestone, Queens, was the first to receive Time Warner Cable's 150-channel Quantum service in December 1991.

Chapter 4

Taking the Lead

1990–1995

In the period leading up to the Time-Warner merger, Warner Cable President James L. Gray, center, assembled a strong regional operating team, including Charles E. Smith, Jr., left, and Frank Nowaczek, right.

Warner Cable president Jim Gray and his management team wanted to make a good first impression on their soon-to-be new boss, ATC chairman and CEO Joe Collins. Shortly after Time Inc. announced its purchase of Warner Communications in 1989, Collins planned a trip to Columbus, Ohio, to meet the Warner Cable team. Gray booked a room at the nearby country club and planned a lavish luncheon. He was investing in the working relationship between the two businesses, after all, and wanted to create the right atmosphere for Collins to get to know the Warner Cable leaders.

Collins rose to the top of ATC on the strength of his operating acumen and skill at matching the right manager with the right job, not by pairing wines with entrée selections. He took one look at the country club menu, turned to the person next to him and asked, "Could I just order a hamburger?"[1]

Collins' comments during the meal tended to focus on fairly straightforward, "meat and potatoes" operational issues as well, recalled Lynn Yaeger, then-Warner Cable's senior vice president, corporate affairs. No inspirational pronouncements of the sort that Steve Ross was so fond of delivering. "After he left," Yaeger said, "the rest of us were all looking at each other saying, 'Well, what's this going to be like ... ?'"[2]

Technology Leader

It wasn't the first, and certainly not the last, time during the first half of the 1990s that the combination of the Time Inc. and Warner businesses and corporate cultures led people from both sides of the deal to wonder, "What's this going to be like?" Indeed, it took years for the two sides to truly come together. At times, the executive-level conflicts resembled *Clash of the Titans*. When it came to integrating the two cable businesses that had been among the key drivers for the takeover, however, things went more smoothly.

From the start, the combined entity assumed its position as cable industry technology leader, which the company arguably retains to this day. The promise of two-way cable technologies launched ahead of their time finally began to be realized. Hybrid fiber-optic architecture was increasingly adopted by Warner Cable and ATC systems across the country. And they learned from the most tumultuous regulatory period in the cable industry's history to work more effectively with regulators and legislators in Washington. As a result, the company emerged by the mid-1990s as the industry bellwether in providing improved customer service.

Dick Aurelio, the former New York City deputy mayor who had played an important role in expanding Warner Cable's franchising activities, returned to New York City to oversee the construction of the Brooklyn-Queens franchise, and was Steve Ross' pick to run the combined Time Warner cable operations in the Big Apple.

Focus on New York City

While Warner Cable and ATC weren't formally combined into a single entity until 1992—in part due to the fact that 18 percent of ATC stock was held by the public when Time Inc. bought Warner—there were opportunities to join forces almost immediately. One of the earliest and highest-profile examples was in New York City, where the combined cable systems served more than 770,000 homes and comprised the "largest contiguous cable operation in the world."[3]

Steve Ross was acutely aware of the fact that Time Inc. and Warner were creating the world's largest communications and entertainment company, headquartered in the communications and media capital. Even before the deal was announced, he wanted to ensure that the company's combined cable systems had the right person running the show in the Big Apple.

Dick Aurelio had returned to New York in 1986, dispirited by the decision to sell many of the prize cable franchises he had helped land for Warner Amex. After a brief stint as head of programming, marketing and sales for Warner Cable, he had told Drew Lewis that he wanted to lead the build-out of the Brooklyn-Queens system, overseeing operations from an office in Flushing. He also had met with Ross to get his blessing.

"Cable was taking a beating generally for lousy customer service," Aurelio told Ross. Warner and most other operators were focused on building out systems and generating revenue as quickly as possible to the exclusion of most other concerns. Taking over the Brooklyn-Queens system would allow Aurelio to return to his beloved New York, as well as "take responsibility for a new cable operation in which I could put my own stamp of customer service first." Indeed, one of the posters on the wall of Aurelio's office reminded employees that "We're not in the cable television business, we're in the customer service business."[4]

For Ross, there was no one better suited for the herculean task of running Time Warner's New York cable operations. Aurelio returned to Ross'

Time Warner Cable formed a joint venture with Queens Inner Unity Cable System, QUICS, the largest minority-owned cable company in the country. Pictured at the system's 1989 groundbreaking ceremonies (left to right) New York City Councilman Archie Spigner, Queens Borough President Claire Shulman, QUICS Chairman Percy E. Sutton, Congressman Floyd H. Flake, and QUICS General Manager Lawrence Jarrett.

office in 1989 to learn the proposed merger of Time Inc. and Warner was going to be announced the following day. Ross told him that he had insisted to his counterparts at Time Inc. that Aurelio, based on his experience and contacts in New York business and political circles, was the best choice to run the combined New York City cable operations of the two companies. Aurelio set about planning how to combine and coordinate operations of the cable systems in Manhattan, Brooklyn, and Queens, even though it would be months before the deal was finalized. He knew he could easily hand the reins for the day-to-day operations of the Queens and Brooklyn systems to Barry Rosenblum, who had impressed him as manager of ATC's smaller Queens system.[5]

After Aurelio became president of the combined New York City cable systems, Jack Gault, who had been president of ATC operations in New York City, transferred to Stamford, Connecticut, where the national headquarters of the combined cable operations was subsequently located, as one of several executive vice presidents under Joe Collins.[6]

"Cable's Finest Hour"

CNN had been steadily gaining an ever-larger viewership as the 1980s progressed, while network television was losing viewers and slashing budgets, often by cutting overseas correspondents and news bureaus. The advent of a new decade and a series of world-changing events unequivocally put CNN, and cable television, at the center of global media. As ATC's 1990 annual report noted:

Never in its history has the promise of cable television come closer to being fulfilled than during 1990 and the early days of 1991. The past year was fraught with political change and social upheaval around the world. There were peaks of exhilaration as we witnessed the resurrection of democracy in Eastern Europe and valleys of apprehension as we watched events unfold in the Persian Gulf.

Ironically, cable's finest hour was the nation's gravest. For the five months following Iraq's August 2 invasion of Kuwait, CNN, the 24-hour news channel, offered comprehensive coverage of the Allied buildup in the Saudi desert. From world capitals, CNN correspondents reported on the diplomatic efforts to end the crisis. Meanwhile, during the critical debates and subsequent votes in the U.S. Senate and House of Representatives, cable viewers were riveted to C-SPAN, cable's public affairs network, as the nation's lawmakers voted to support the United Nations resolution approving the use of force to oust the invaders from Kuwait.

As the Middle East teetered on the brink of war and finally erupted, CNN became the network of choice for political and military leaders of both sides, as well as millions of concerned viewers everywhere.[7]

CNN anchor Bernard Shaw's reporting during the first Gulf War helped make CNN the default news source of choice for millions of television viewers around the globe in 1990 and 1991.

CNN correspondent Peter Arnett's reports from the streets of Baghdad during the first Gulf War gave global viewers a rare glimpse of the war's real-time impact. They also sparked criticism that CNN's reporting was tightly constrained by Iraqi officials.

Quantum

Chief technology officer Jim Chiddix was rehearsing comments he was planning as part of a larger Time Warner management presentation in the months following the merger. Assembled in the old Warner Bros. boardroom awaiting their turns at the podium were Steve Ross, Jerry Levin, Joe Collins, and others. "My dry run was the story of how fiber was transforming cable into something very different, a communications network with the potential for interactivity and the ability to ... transmit 1,000 megahertz of spectrum to our customers," Chiddix recalled.[8]

"Now, Steve was the same guy who'd launched QUBE; he loved new technologies, and he began really paying attention to what I was saying and then he stopped me and said, 'Now let me get this straight. You mean this technology exists today? This fiber stuff? You could really do this and deliver a gigahertz of services?'" Chiddix assured him that the technology existed. "Well, by George, we're going to do this. You go out and build this in one of our cable systems," Ross said.[9]

Levin wanted the test built in the New York City area so the company could showcase it for Wall Street analysts. Sections of the new cable system being built in Queens were chosen as a test bed for what at the time was a state-of-the-art 150-channel system that offered more than twice the capacity of most systems around the country. The systems were being built out using optical fiber for the trunk lines and coaxial cable to make the "last mile" connection to individual homes.[10]

Fiber-optic cable was used in Queens, New York, in 1991 to provide an early showcase for the new technology's ability to deliver a 150-channel cable system dubbed Quantum.

Veteran Warner Cable engineer Roosevelt "Rosey" Mikhail provided the technical solution that eliminated signal interference and enabled Time Warner Cable's NY 1, a 24-hour cable news channel, to be transmitted on channel one on the cable system.

"It's Amazing"

In March 1991, the company went public with its plans for what Nick Nicholas, who succeeded Munro as co-CEO in January of that year, heralded as "the world's first 150-channel, two-way interactive cable TV system."[11] Dubbed Quantum, the system, planned initially for the Whitestone and Bellerose neighborhoods in Queens, was "leading the way toward the long-anticipated video and information pathway into the home of the next century." Interactivity, including 57 channels dedicated to pay-per-view movies, was the key to the new system, Nicholas said.[12]

Nicholas noted that Warner Cable's QUBE had been the world's first interactive system, but "QUBE was ahead of its time. Now, for Time Warner Cable subscribers, interactivity's time has come," he declared.[13] Jerry Levin added that the company was achieving these best-in-class benchmarks with Quantum despite the fact that it was still using analog signals. Converting analog signals to digital ones and zeros and compressing them to achieve greater bandwidth was on the drawing boards at Warner Cable and other leading cable operators, but had not been deployed on a commercial basis as of 1991.

Quantum went live on December 18, 1991. The Massa family of Whitestone, Queens, existing subscribers to the company's Brooklyn-Queens Cable system, inaugurated the new service. "We're thrilled to be the first with Quantum," said Lori Massa. "You press a button and get an immediate choice of 15 movies at your fingertips, with neat displays right on the screen. It's amazing."[14]

Barry Rosenblum, who was running the Queens operation, noted that the Massas were not alone. "Buy rates went up significantly" thanks to the ability to order movies with start times in half-hour increments. "People loved it," he said. The system's operations room full of 75 to 80 automated tape decks starting movies on the hour and half-hour was hardly cutting-edge technology itself, Rosenblum conceded. But Quantum gave the Warner Cable team confidence that consumer demand would justify offering video on demand—the Holy Grail of cable technologists for years—as soon as advances in cable technology made that possible, even on a test basis.[15] In less than two years' time the high-tech team reassembled, this time in Orlando, to once again test the limits of the possible in cable technology.

NY 1 News

While Rosenblum was rolling out the state-of-the-art cable service in Queens, Aurelio was in his Rockefeller Center office planning another challenge to the cable industry status quo. As early as his 1989 meeting with Ross on the eve of the merger announcement, Aurelio had broached an idea he had been developing that would help define the combined cable companies' presence in New York. It would also differentiate the company from competing fare offered by satellite television and, sooner or later, telephone companies.

Aurelio wanted to launch a 24-hour cable news channel just for the New York market. "I thought that would give us a terrific edge with the competition, which was going to come inevitably at some point," Aurelio recalled. Ross was on board immediately. "He said, 'That kind of vision is exactly why I want you there. I will support that. You just tell me what we have to do and when we have to do it, and I'll support you on that.'"[16]

Bringing the instincts of a marketer, not an engineer, to the project proved to be a key to its success. Aurelio wanted to call the channel NY 1—"because there's only one New York"—and he wanted it to be carried on channel one in New York City. Trouble was, according to every engineer he talked to, it wasn't practical to carry a television signal on channel one—too much signal interference. Indeed, virtually every cable system in the country, not to mention broadcasters, left channel one vacant on their systems.

Aurelio turned to the Warner Cable "engineering guru" he had tapped to help design the Brooklyn-Queens system—Roosevelt "Rosey" Mikhail. "Rosey, you've got to find a way for me to use channel one. I don't know how you do it, but you've got to figure out a way to do it," he pleaded. "I know no one else in the country is doing this, but we've got to do it because I would like this channel, this new channel, to be called NY 1," Aurelio said. Mikhail said that he would think about it. Within 24 hours, Mikhail, who had studied engineering in his native Egypt before moving to the United States, determined that he could make some rather minor adjustments to eliminate the interference and make channel one fully functional.[17]

NY 1 News went live on September 8, 1992, and was a success almost from the moment it launched. The award-winning 24-hour news service rapidly became an important feature in local political coverage and as a watchdog monitoring municipal government machinations. And it became the model for a number of 24-hour news stations in major markets served by cable operators, including several more launched by Time Warner Cable.

"Zapped!"

The theft of cable television signals had become a multi-million-dollar annual cost of doing business for the increasingly urban cable television industry by the early 1990s. Tens of thousands of subscribers across the country were either stealing signals outright by splicing into existing cables, or tampering with cable converter boxes to steal premium channels like HBO. Barry Rosenblum, president of American Cablevision of Queens (ACQ), the Time Warner subsidiary operating in the New York City borough, decided in early 1991 that enough was enough. It was time to fight back.[18]

Working with its addressable converter box manufacturers, ACQ devised an "electronic bullet." It took the form of a signal that could be sent to subscribers that instantly scrambled the signal if the box had been tampered with, while sparing boxes that had not been altered. Rosenblum's engineers pulled the "trigger" shortly after 7:30 p.m. on March 13, 1991.

The shot heard 'round the borough—and 'round the country thanks to widespread media coverage once the company publicly revealed the sting operation the following month—led more than 300 unsuspecting ACQ customers to call the company's service phone line demanding that their television signal be restored. The company hit them with a federal lawsuit alleging theft instead. The front page of the *New York Post*'s April 25, 1991, edition carried an image of a television with a snowy image of static, and one word emblazoned across the picture in capital letters: ZAPPED!

Rosenblum let it be known that this was only the first shot in a long-term war against signal theft, and that other operators around the country were sure to be adopting similar anti-theft steps as soon as possible. This was no victimless crime. As with any business, the cost of theft is inevitably passed on to the consumer.

Becoming Time Warner Cable

After more than a year of operating Warner Cable and ATC as separate entities, with ATC's Collins clearly in charge, Jim Gray decided it was time to take the operating relationship to the next level. No fancy lunch this time. Gray called Collins and said he wanted to come to Stamford and talk about the future. "I knew if we combined the organizations there were going to be some significant savings and improvements. In addition, my people needed to have a chance to compete, and the only way they could compete would be to get in the action," said Gray.[19]

Collins agreed, and the ATC and Warner Cable systems were combined in 1992. Collins named Gray vice chairman of Time Warner Cable. When it came to staffing management of the systems, Collins' marching orders to Gray and Time Warner Cable president Jimmy Doolittle were straightforward: No playing favorites. "Joe had a very clear view of the fact that one company bought the other; that's not what's important here. What's important is we want to put the best people in all of the key slots that we can from whatever their pedigree," said Gray, who was intensely proud of the legacy his team at Warner Cable had created.[20] "I was coming up toward retirement, and I just wanted it to be put in a good position so that what we had worked so hard for would have some meaning,"[21] said Gray, who, after two decades with Warner Cable and then Time Warner Cable, retired in 1994. He didn't retire from the industry, however. The following year, Gray agreed to serve as chairman and CEO of PrimeStar, the jointly owned cable industry satellite venture, and stepped down from that position in 1997.

Terry O'Connell, Carol Hevey, Kevin Leddy, Gary Matz, and a host of former Warner Cable executives who continued to rise through the ranks at Time Warner Cable are indebted to Gray for his personal commitment to seeing that their careers continued full steam ahead. Gray "accepted the role of vice chair and actually rented an apartment up in Connecticut," noted O'Connell. "A lot of the old Warner Cable executives knew that one of the primary reasons Jim did that was to make sure his team integrated well into the acquiring cable executive group and company," he added.[22]

Reducing Debt, Increasing Drama

The early successes scored at the operating level by the combined ATC and Warner Cable systems in New York City as well as across the country took place against a backdrop of often unsettling executive suite drama. One of the top priorities of the executive management team at Time Warner was to reduce the $11.2 billion mountain of debt amassed to finance Time Inc.'s purchase of Warner Communications. The hoped-for synergies resulting from combining forces and riding the wave of globalization in business, a subject that dominated most business discussions in the early 1990s, were not expected to reach fruition until Time Warner had its balance sheet under control.

Time Warner pursued numerous options to both raise capital in the public markets and attract large corporate investors interested in buying a stake in their business, in one form or another. The Japanese trading company C. Itoh expressed an interest in early 1990 in forming a partnership with Time Warner. Toshiba Corp. also engaged in

The 1992 combination of the management and operations of ATC and Warner Cable was big news to the employees of both companies, which had been managed separately even though they had been operating under the same corporate umbrella since the merger of Time Inc. and Warner Communications.

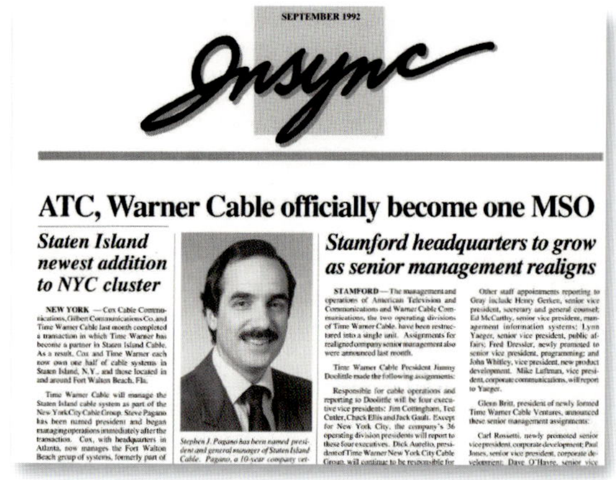
SEPTEMBER 1992
Insync
ATC, Warner Cable officially become one MSO
Staten Island newest addition to NYC cluster
Stamford headquarters to grow as senior management realigns

Now we are one!
AUGUST/SEPTEMBER, 1992
Warner Cable News
Published for employees of Warner Cable Communications Inc.
Here's who's who in the new top TimeWarner management staff
Warner Cable and ATC merged; become one TimeWarner Cable
Warner's Dublin Corporate Headquarters to close; Jim Gray named vice chairman

discussions with Time Warner during this period. Japanese stock and real estate markets had attained stratospheric levels, and Japanese investors were buying so-called trophy assets in the States and Europe, including Rockefeller Center in New York and Columbia Pictures and MCA in Los Angeles. Time Warner officials also held numerous talks in 1990 and 1991 with other potential investors in Japan and Europe. No potential deal progressed beyond the talking stage, however.

One likely reason for investor inaction was the debt burden hanging over Time Warner. Steve Ross and close advisors from the Warner side of the company advocated a complex rights offering—meaning only existing shareholders had the right to buy stock being offered—as a means of raising $3 billion or more in capital to retire a significant portion of the debt. Announced in June 1991, the deal was seen as so detrimental to existing shareholders that Time Warner's stock price plunged nearly 25 percent in a matter of days to slightly more than $88 a share.[23]

Major shareholders were incensed. Time Inc. directors had rejected the $200 a share offer from Paramount in 1989 in part by arguing that the stock price of the combined companies would most likely trade well north of that amount in a few years' time. Here it was a few years later and the stock was trading at less than half the Paramount offering price. To make matters worse, the rights offering hit the newswires shortly after it had been disclosed that Steve Ross' total compensation for 1990 was a staggering $78.2 million—$74.9 million of which was defined as representing his financial interest as a founder of Warner. Tens of millions in additional compensation were due Ross over the next few years under the terms of his Time Warner contract.[24]

Time Warner determined that it had to appease angry shareholders and regulators who had indicated they were opposed to the rights offering as originally configured. Time Warner simplified the structure of the transaction, In part at the instigation of co-CEO Nick Nicholas, and in early August 1991 raised $2.76 billion. Now it was time to redouble efforts to find some international investors willing to partner with Time Warner.[25]

The initial post-merger Time Warner management team of co-CEOs Nick Nicholas, left, and Steve Ross, right, flanking chairman Dick Munro, never effectively worked together.

Ross-Nicholas Rift

With months passing and the company unable to deliver on most of its promised synergies, tensions grew between co-CEOs Ross and Nicholas. Nicholas had been advocating a straightforward sale of non-core assets to pare down debt—a strategy that would have been familiar to any business school graduate. He even floated the idea of selling Warner's music business—not that he thought it was an unimportant business, but on the grounds that it could fetch billions on its own—or create a music joint venture with another industry leader. "I was saying, 'What businesses do we want Time Warner to be in going forward? Doesn't mean it's a bad business,'" Nicholas later recalled.[26]

Jerry Levin, who was named co-CEO of Time Warner after the board ousted Nick Nicholas from that position in early 1992, became the sole CEO and chairman following the death of Steve Ross in December 1992.

Ross and his confidantes, including longtime advisor Oded "Ed" Aboodi, favored a different path. They wanted to construct a partnership, Time Warner Entertainment Company, L.P. (TWE), which would include several corporate crown jewels, namely the cable businesses, Warner Bros., and Home Box Office. They then would seek partners willing to invest in the new unit in return for a minority equity stake. A key aspect of the transaction would be the transfer of more than $6 billion of debt off of Time Warner's books and onto the balance sheet of the TWE partnership.[27]

With the rights offering behind them, Time Warner executives were able to rekindle talks with C. Itoh and Toshiba. A deal was proposed and announced in October 1991 in which both companies would invest $500 million each in Time Warner Entertainment in return for a combined 12.5 percent ownership stake. Nicholas scoffed at the idea of selling a portion of these businesses' upside and future growth for a mere $1 billion. Earlier in the year, Time Warner executives had said repeatedly that they expected to get $2 billion from these or other investors.[28]

Another reason for wanting to bring in Japanese investment partners, noted Robert Marcus, who had been working on Time Warner issues almost from the moment he joined the New York law firm of Paul, Weiss, Rifkind, Wharton and Garrison right out of law school in 1990, was to create satellite partnerships in their home country. "We created TWE Japan, which was kind of a parallel structure," he said. At a later date, TWE Japan did go into the cable business in Japan, but the initial focus was on selling Warner Bros. products in Japan.[29] For much of the previous year, Jerry Levin had worked closely with Ross and Aboodi. That relationship was made official in May 1991 when Levin added the title of Time Warner chief operating officer to that of vice chairman. The promotion was seen as having the effect of distancing Nicholas from Time Warner's day-to-day operations and senior managers.

With co-CEOs Nicholas and Ross increasingly at odds over TWE and other issues as 1991 progressed (though Nicholas eventually acquiesced to the creation of TWE), sentiment began to spread among certain Time Warner executives and board members that Levin should replace Nicholas. The succession issue was given an added sense of urgency in November of that year when Ross confirmed what Munro, Nicholas, and others had privately learned—that he had prostate cancer. Worse, he was suffering a recurrence and was undergoing chemotherapy.[30]

By January 1992, Ross and his Warner supporters, joined by vice chairman and COO Jerry Levin, Michael Fuchs of HBO, and others from the Time Inc. side, set in motion a series of steps that would lead to Nicholas' ouster. Events came to a head in February as Nicholas and his family were on their annual ski vacation in Vail, Colorado. Board members were discretely contacted and told that Ross could no longer work with Nicholas and that they should support replacing Nicholas with Levin.

By the time Nicholas was notified that a board meeting was going to be held the following day to consider his fate, it was a done deal. He refused to return to New York for the meeting and insisted that the board would have to fire him; he wasn't about to resign. In the end, the sole director who declined to vote for his ouster was the last heir of the founding family of Time Inc. to serve on the company board—Henry Luce III.[31]

Ross' executive suite victory was quite literally short-lived. Tragically, he was losing a much more important battle with cancer. His co-CEO, Levin, took on an increasing amount of the CEO duties as 1992 progressed and Ross' health declined. Failing in a final attempt to stanch the spread of his cancer, Ross died in Los Angeles in December 1992. Levin was named chairman and CEO.

Steven J. Ross, 1927-1992

Steven J. Ross, 1990.

Steve Ross' health continued to deteriorate throughout 1992. He retained the title of co-CEO, but his counterpart, Jerry Levin, assumed much of the chief executive duties. Ross died on December 20, 1992, in Los Angeles, weeks after having undergone surgery as a last-ditch attempt to halt his cancer's progress.

Despite criticism of Ross because of Time Warner's difficulties in moving forward as a combined entity and Ross' enormous payout from the sale of Warner to Time Inc., the troubles did not overshadow a career's worth of outsized achievement.

Although Warner reportedly spent lavishly on corporate clients and executives during Ross' watch, he also created a tremendous amount of wealth for shareholders over his decades in business. When he took his small limousine and rental car business public in 1962, it had a market value of $12.5 million. He sold its ultimate successor, Warner Communications, to Time Inc. for more than 1,000 times that amount, or $14 billion.

His vision and risk-taking were legendary; Ross' investments in cable television being a case in point. As his obituary in *The New York Times* noted, "He gambled on the music video channel MTV and on the Nickelodeon cable service of children's programming when most other corporate leaders thought cable was just a piece of thick wire."[32]

Steve Ross continues to be fondly remembered by those who worked for him as an inspirational leader committed to bringing out the best in those he entrusted with building the organization.

Despite the initial handshakes and smiles between Time Warner chairman and CEO Jerry Levin and US West CEO Richard McCormick, US West's purchase of a 25.51 percent stake in the TWE partnership in 1993 led to years of often tense relations between the Bell operating company and Time Warner.

US West Calling

Even before the TWE partnership was capitalized in June 1992 with the contributions from Time Warner as well as its Japanese partners, Time Warner executives were searching for additional equity investors. The goal was to find someone to help finance the continued growth of the cable and other businesses in the partnership, as well as to bring in additional management expertise. The regional Bell operating companies (RBOCs) formed from the dismantling of AT&T were obvious candidates.

"We also, around that time, realized," said Glenn Britt, then the president of Time Warner Cable Ventures, "that our plant, if we did the right things with it, could be in the phone business," as demonstrated with the Quantum rollout in Queens. "We said, 'Well, that's great, but we don't have a clue how to be in the phone business, so it wouldn't be bad to have a partner.'"[33]

Time Warner initially talked with several phone companies and anticipated having two to three of them make investments equal to the investments made by Toshiba and C. Itoh, which shortly after TWE was formed changed its name to Itochu. US West, based in Denver, "said that they wanted to be the only one," noted Britt. "They anted up $2.55 billion for that privilege, so we ended up with them as partners."[34] The addition of US West to the TWE roster of partners was finalized on September 15, 1993, giving the bell operating company a 25.51 percent stake in the partnership.[35]

"A Less Than Harmonious Partnership"

In theory, a partnership between a cable company and phone company ought to have functioned well. First, AT&T and then its corporate offspring had wanted to carry television signals for years. Efforts gathering momentum in Congress and the courts suggested that it was a question of when, not if, this was going to become a reality. In the fall of 1993, TCI announced plans to merge with Bell Atlantic, and Cox Communications announced a deal with Southwestern Bell.[36] Both deals fell through the following year, but the intersection of the two industries seemed all but inevitable.

In practice, the US West investment in TWE generated a great deal of static. Time Warner bore a share of the blame, noted Marcus. US West owned slightly more than one-quarter of the partnership, and Time Warner was designated as the manager of the cable systems. "But as it related to the cable business, we actually created a governance structure where, on a lot of matters, it was 50/50 ... that was, I think, certainly in hindsight, a recipe for a less than harmonious partnership."[37]

US West's actions didn't improve the relationship. "They had great aspirations to be in the cable business in a bigger way than what the investment in Time Warner Entertainment enabled them to do. So there were consistently issues of them trying to do investments outside of the partnership, frequently coming to us for consent," Marcus said. "There was misalignment almost from day one. And the cultures were very different," he added.[38]

The phone company still operated with a utility company mindset. It was heavy on layers of management and light on innovation, noted Joe Collins. "They had more people in their headquarters in Denver monitoring their investment in the cable company than we had at the headquarters of the cable company running the whole business," he said.[39]

The Full Service Network

Time Warner Cable's ranking in the vanguard of cable industry innovators played a key role in US West's decision to invest $2.55 billion in the TWE partnership. Specifically, the phone company was excited by Time Warner Cable's plans, made public in January 1993, to develop its Full Service Network (FSN)—the nation's first truly digital broadband interactive cable television system.[40]

Indeed, the TWE agreement that US West signed in September of that year provided that TWE would begin to implement some aspects of the FSN technology over the next five years.[41]

Jerry Levin introduced Time Warner Cable's Full Service Network in Orlando in 1994. As a broadband technology test bed, the FSN was a success, even if many of the applications were not yet economically or commercially feasible.

Although Time Warner Cable did develop and roll out the required core FSN technology, deployment of all the FSN bells and whistles never became a reality. Yet the digital network created by that technology clearly enticed the phone company to join the communications future.

Experience it. Use your smartphone to watch a video of Jerry Levin describing the concepts behind the Full Service Network that later drove video on demand and related services, which gave customers greater convenience and control.

The End of a Standalone ATC

In order to include ATC in the Time Warner Entertainment partnership, Time Warner needed to buy out the remaining public shareholders of ATC's common stock. On June 26, 1992, the date the TWE partnership closed, ATC once again reverted to private ownership, just as it had in 1978 when Time Inc. bought out the remaining public shareholders to own 100 percent of the cable operator. Time Warner acquired the 18.7 percent interest in ATC that it didn't already own from public shareholders. ATC and Warner Cable had already been working together under a joint operating agreement as Time Warner Cable. The buyout marked the official end of ATC as a standalone corporation.[42]

The Full Service Network demonstrated many "Home of the Future" broadband technologies, such as video on demand, which Time Warner Cable helped make commonplace in the years ahead.

The significance of the FSN extended far beyond its role in drawing US West into the TWE partnership. In addition to functioning as a proving ground for next-generation cable technology, the FSN employed many of the technologies and concepts that in short order would be central to the popularization of World Wide Web browsing technology. That technology transformed the Internet from a computer-based communications system used mostly by academics and scientists to one linking individuals the world over. In fact, Jim Clark, chairman of Silicon Graphics, one of the vendors Time Warner Cable brought in to help design aspects of the FSN, left to form Netscape, a pioneer in Web browsing technology. And software developed initially for the FSN by Sun Microsystems, but not chosen for the project, evolved into the widely used Java computer programming language.[43]

"Let's Build a Time Machine"

The FSN grew out of a meeting held by Time Warner Cable's senior management team of Collins, Doolittle, and Britt with then–Time Warner co-CEO Levin in the fall of 1992. They were discussing the next year's budget for the cable business when talk turned to the technology horizon. Quantum was impressive, especially in its use of hybrid fiber coax technology, but still reflected the cable industry's analog signal past. Digital was clearly the next step. A host of other companies, from AT&T to IBM, were reportedly considering similar moves. TCI's John Malone was telling the public that digital compression would enable his systems to deliver 500 channels without the need for costly fiber-optic upgrades.[44]

Guiding Light

Even as the FSN was still being developed, Chiddix and the Time Warner Cable engineering team in 1994 received an Emmy® Award, one of television's top honors, for "developing and advocating broad band fiber-optic technology and architectures." It represented the much-deserved recognition from the industry's top engineers and technology gurus of what the Time Warner Cable fiber-optics team had accomplished and contributed to the industry.

The Time Warner Cable team had reached the pinnacle of achievement in their field, but like many communications technicians, realized that their contributions would remain largely invisible to the viewing public. This point was brought home to Chiddix, who had made the acceptance speech on behalf of the group, in humorous fashion as he exited Manhattan's Marriott Marquis Hotel, where the award ceremony was held. "I'm walking through Times Square carrying an Emmy statue and people began coming over and asking which soap opera I was on."[45]

Time Warner Cable's engineering team won an Emmy® Award in 1994 for its pioneering work in broadband fiber-optic technology.

Costing $4,500 each and resembling the size of a small refrigerator, the home communications terminal used in the homes of subscribers to the Full Service Network clearly were not ready for commercial production.

Jim Chiddix and his team had made it clear that all the pieces were in place for digital television, video on demand, and two-way interactivity. "By this time we had done lots of demonstrations, and I had given lots of speeches, and others had as well, about the future of cable and where it was going with the combination of this hybrid fiber coax distribution architecture and digital technology—digital television technology," Chiddix said.[46]

"As soon as you ran fiber to different neighborhoods, you could send different channels to different neighborhoods. And the implications of that were that because you had few enough homes in a neighborhood, you could actually send an individual channel to an individual home. Because video used so much bandwidth, the idea of sending an individual video program to an individual user had been unthinkable before that. But that made it possible to begin to think about video on demand, delivering a program to a home. And that was kind of a revolutionary thought."[47]

The question was how to take those advances into the real world and see how customers reacted. Britt said, "You know, we ought to take a few thousand homes somewhere and do like the ultimate, whatever we think we could do."[48] Collins used one of Chiddix's favorite phrases to describe the approach. "Let's build a time machine," he said. "Let's take a cable system and pretend that we can look out there and get 10-years-from-now technology and have it be working at a useful basis, and bring it back and build it, and then we'll use it to see what we can deliver on this two-way plant."[49] The group assigned Britt and Doolittle the task of drawing up a budget for the project. "We came up with a number, which I can't remember," Britt said, "but it was a lot less than we ended up spending."[50]

Great Expectations

The cable executive leadership and Chiddix clearly saw the FSN as a limited consumer test of cutting-edge technology. Once again, the company returned to the suburbs of Orlando, Florida, to test its latest capabilities. The goal was to demonstrate that the technology functioned in the field and to gauge customer interest in the products and services that the melding of hybrid fiber coax systems and digital could offer, especially video on demand. Jerry Levin had a different idea.

Levin, who had been named Time Warner's chairman and sole CEO following Steve Ross' death in December 1992, unveiled plans for the FSN on January 26, 1993. He described the dawning of a "brave new world" with the creation of an "electronic superhighway" in suburban Orlando, scheduled to go live in April 1994. "I don't want to wait for the 21st century; we're going ahead and building," he told the media. "We aren't waiting for some test results. ..."[51] The *Wall Street Journal* reported that Levin "indicated that Time Warner itself (was) talking with potential partners in the communications and consumer electronics industries to help launch its systems nationwide and overseas."[52]

Mike Hayashi, who had joined Chiddix's team at Time Warner Cable from set-top box developer Scientific-Atlanta shortly before the FSN was announced, said the April 1994 target launch was a "crazy date. All I remember is that the cable industry was barely learning how to use these advanced analog set-top boxes, and we have this incredible experiment that's been announced that promised the future."[53] In fact, the launch date was delayed, and eventually set for December 1994.

Mike Hayashi, now Time Warner Cable executive vice president of architecture, development and engineering, joined the company's technology group in 1992 from set-top box developer Scientific-Atlanta and immediately went to work on developing digital set-top box technology for the FSN.

"The Reality Was the Complexity"

With expectations set so high, it was all but inevitable that the fledgling effort would disappoint in one aspect or another. As it turned out, Time Warner Cable and its partners, including Silicon Graphics, Scientific-Atlanta, and AT&T's Network Systems division, which was later spun off as Lucent Technologies, greatly underestimated the challenges they faced in developing the FSN and working effectively together as part of such an unprecedented enterprise.

Carl Rossetti, senior vice president of corporate development at the time, had direct oversight of the effort. "What the attraction was from a business point of view was the simplicity," he said. "The reality was the complexity. And if anyone had thought of what this would cost ahead of time no one would have ever done it."[54] Take the home communications terminal inside each of the 4,000 FSN subscribers' homes. Each unit cost roughly $4,500, was the size of small refrigerator, required a fan to keep it cool, and was housed in a custom-designed piece of furniture.

Lessons Learned

The development headaches suffered by the FSN team were soon forgotten, as Jerry Levin and Jim Chiddix used a TV remote to launch the network on December 14, 1994, in Orlando. An image of a carousel navigator appeared on a projection screen, and Levin demonstrated the first-ever digitally streamed movie over a cable system. "After all of the nightmares, and all of the battles and negotiations and the long days and the traveling, to have that music come up and the carousel start, it was like you went unconscious almost. That was spine-*ting*ling. It was huge; it was huge," said Jim Ludington, who had been the first employee officially designated as working on the FSN.[55] When the FSN was finally completed, Ludington made the first television shopping purchase from a home he had bought, which was part of the test neighborhood in Orlando: He bought a Looney Tunes cookie jar.[56]

Joe Collins conceded that the FSN was vastly oversold and over budget. Yet in terms of providing real-time data on what customers did and didn't like about video on demand and interactivity, it was a genuine success. "It was really always an experiment," he said. "We knew this stuff would never scale, but it was a giant marketing test to find out what parts of this technology were really saleable by the customers and what they would buy."[57] Time Warner Cable tweaked the test in Orlando as it gauged consumer interest over the following months. By the time the company announced in April 1997 that it was phasing out the FSN, many of the advances that had been pioneered in suburban Orlando were already being incorporated in next-generation set-top boxes and other cable products and services.[58]

1992 Cable Act

Consumers were benefiting from Time Warner Cable's continuing advances in applying hybrid fiber and digital technology to its nationwide cable plant during the early 1990s. To a lesser extent, other cable operators were upgrading their technology as well. Yet the general public and their representatives in Washington, D.C., did not associate cable companies with technological bells and whistles. Two issues were top of mind with consumers when it came to their cable television provider: high prices and poor customer service.

Despite the warnings of Trygve Myhren and Bill Daniels in the mid-1980s, few cable companies had shown much restraint when it came to upping prices following the passage of the '84 Act, which largely deregulated the industry. Shortly after Myhren resigned in 1988, ATC raised prices significantly in Hawaii, even though Myhren had a tacit understanding with regulators that rates would increase only modestly.[59] Similar stories played out across the country.

ATC had already felt the heat as the 1990s began. It had been part of an industry-wide effort to improve customer service standards in 1990, "but more needs to be accomplished," its annual report for that year noted.[60] The report said that legislation to reregulate the industry was debated in both houses of Congress that year, and that further hearings were scheduled for 1991. The FCC was also examining the industry to see whether regulatory changes were in order. The company let its shareholders know that management was often in D.C. pleading the company's case, but wasn't placing odds on the outcome.[61]

After passage of the 1984 Act, "The industry as a whole did not act like you need to act if you're a regulated business, which is participating in the political process and the fund-raising and lobbying and all this stuff that goes on," Britt noted. "So the broadcasters exploited that vulnerability and created this notion that cable was an unregulated monopoly and was bad, and that free TV was going to be destroyed," he added. It's worth recalling that until the early 1990s, "more than half of the homes in the country still got their television off-air, and there really wasn't any competition to cable, so off-air TV was pretty important."[62]

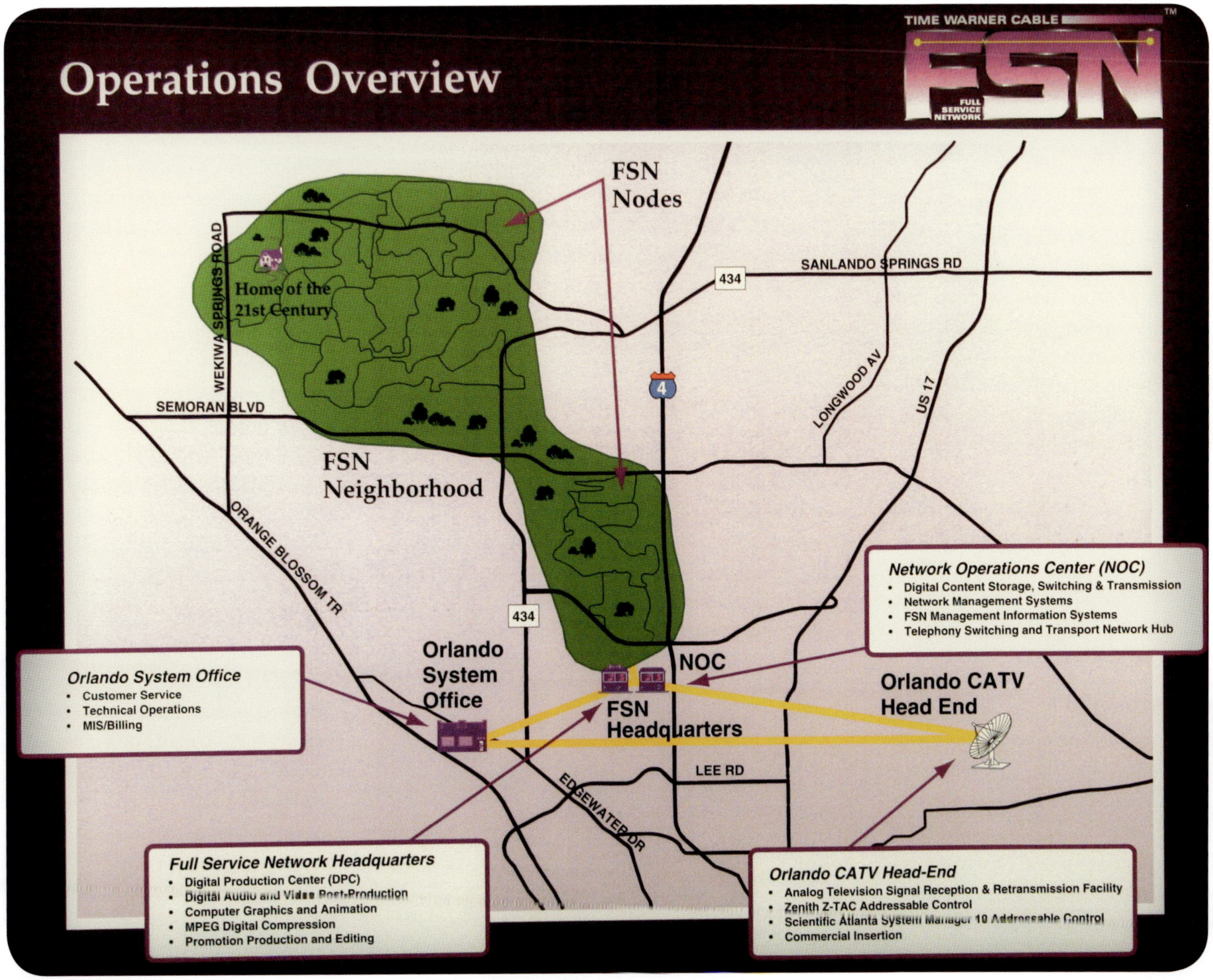

The simple schematic drawing of Full Service Network facilities and subscriber neighborhoods in Orlando belied the tremendous complexity and cost involved in making the technology experiment an operating reality.

Retransmission Consent

The cable television industry grilling increased in intensity and frequency in 1991 and for much of 1992. It culminated in the enactment—over President George H. W. Bush's veto—on October 5, 1992, of the Cable Television Consumer Protection and Competition Act of 1992. Two of the main provisions effectively lowered rates operators could charge subscribers by a total of 17 percent as of May 1994, and required operators to negotiate "retransmission consent" with most television broadcast stations.[63] This retransmission consent was the latest fallout from so-called "must-carry" rules that had been a periodic thorn in the cable industry's side dating back to the 1960s. As a result of the '92 Act, broadcasters could choose either must-carry or retransmission consent for their stations on local cable lineups. Those who selected retransmission consent would then negotiate carriage terms with cable operators. Despite vast changes since then in ownership of cable and broadcast networks, and a communications landscape dramatically different from the one fueling the '92 Act, retransmission consent rules remain in effect with little change, which continues to plague cable systems operators and their customers. The Act also required vertically integrated cable companies that produced programming, like Time Warner, to provide their programming to satellite cable television companies as well.

Time Warner Cable was one of the major parties bringing judicial challenges to major portions of the Act, including its must-carry provisions. Although Time Warner Cable and other challengers lost on the must-carry issue itself twice at the U.S. Supreme Court, the court did recognize that, as a general matter, cable operators are entitled to full-fledged First Amendment protections. In addition, in subsequent related cases, Time Warner Cable was successful in striking down other limitations of cable operators in the Cable Act, including the amount of their own programming they could carry and the number of customers they could reach.

The impact of the '92 Act was felt almost immediately. Time Warner said that its 1993 revenues were reduced by $90 million to $100 million. Time Warner Cable president Jimmy Doolittle said, "That was the only year that I can remember in the company that we did not have at least a double-digit cash flow growth."[64] Another consequence of the legislation was to slam the door on outside investment in the domestic cable industry for a number of years. At the same time, U.S. cable companies and others invested actively in overseas cable companies in search of more favorable regulatory climates.

Rebuilding Public Trust

Time Warner Cable's New York City region was extremely profitable by the time the '92 Act was passed, but it had serious service problems and an historically poor reputation in its Manhattan systems, which was hampered by an older plant that eventually had to be rebuilt. Stories of service crews never showing up or making customers wait all day were legion, even if not always documented as fact. Liberty Cable, an "overbuilder" that competed against Time Warner Cable, was taking out small classified ads at the bottom of the front page of *The New York Times* that listed a building address and declared "another building free from the Time Warner monopoly."[65]

Lynn Yaeger realized by early 1993 that if the company wanted any help from Washington, it would first have to help itself. New York City division president Dick Aurelio added his concern that the New York system would be offering upgraded services and products over the next several years. How was he going to sell these to a public that had such a low opinion of the service he was offering?

Yaeger turned to the communications firm SS+K (Shepardson, Stern, and Kaminsky), which she had first used in the fight against the retransmission consent section of the '92 Act.[66] Their research at that point had led the company to conclude that no matter how logical its argument, it could never win going toe to toe with broadcasters on issues such as who ultimately bears the cost of retransmission consent payments. Cable's reputation with customers was broken almost to the point of no repair. That meant that Time Warner Cable not only needed to change its message, it needed to change its culture if it was going to have a chance of getting the customer back on its side.

On-Time Guarantee

SS+K proposed a sweeping marketing and image campaign that would help rebrand Time Warner Cable in the public's eye. Forget empty platitudes. The public thought all cable companies cared about was money, so the campaign went right for the pocketbook. Time Warner Cable and other major cable companies declared an on-time guarantee. If service people were late for an installation or a service call, it was free.

The company leadership realized that employees from senior management to frontline service personnel were going to have to get behind the initiative if it was going to have a chance of succeeding. Lenny Stern, SS+K principle, noted:

I think we all realized, this was not an ad campaign. This was a cultural change. We spent the first five months going into the field, training employees, having rallies, and making them part of it.

The employees said, "You know, I do my job. We work hard. There are challenges. But I've never been able to look at a customer and have the tool to say, 'Look, hold my feet to the fire, and if we're not here, we're going to take care of you.'"

This on-time guarantee not only mobilized the troops, but it also [enabled] them to have a conversation with consumers about how their behavior was going to change and how the company really meant it.[67]

The Cable Guy

Jim Carrey, rear, with George Segal and Diane Baker, in a scene from 1996's The Cable Guy. *The movie tapped consumer discontent with too-frequent sub-standard levels of cable television service.*

The black comedy *The Cable Guy* starring comedian Jim Carrey did not hit movie theaters until 1996, but its message reflected the widespread consumer antipathy toward cable companies and their employees during the early '90s. The frustration felt by thousands of customers as they waited hours for cable service technicians to show up was a staple of stand-up and late-night comedians as well.

It was not an easy period to be identified as someone working for a cable television operator. Forget the hours spent on late-night calls or repairing storm damage that no one ever heard about. All it took was for someone's neighbor to wait an hour for a service call and all technicians were equally to blame.

Social Contract

While Time Warner Cable was leading the industry in trying to repair its relationship with its customers, it also took the initiative in fence-mending with regulators in Washington in the wake of the '92 Act. To resolve outstanding rate complaints, and to be able to charge rates justified by technological improvements to its systems, Time Warner Cable composed a company-wide social contract rather than set rates on a system-by-system basis according to a complex regulatory framework that was required until most of the '92 Act's rate provisions were repealed by amendments in 1996.

"I went down to the FCC and met with the head of the cable bureau at that time," recalled Jimmy Doolittle. "We basically negotiated a company-wide, instead of doing it on a system-by-system basis, commitment, within five years, to upgrade every system at Time Warner Cable."[68] The cost approached $5 billion by the end of the decade. It was a steep price to pay, but it had the effect of enabling Time Warner Cable to once again plan for future growth with a high degree of certainty.

Time Warner Cable's efforts, and industry-wide actions it helped sponsor, began to repair relations with legislators and regulators. Within a few years, the company had helped pave the way for Congress to take a much more pro-industry approach to telecommunications regulation. That ushered in a new era of growth and innovation that dramatically transformed the company once again.

The Time Warner Cable Technology Center in Manhattan's Harlem neighborhood offers extracurricular opportunities in STEM—science, technology, engineering, and math—education.

Encouraging Education

Time Warner Cable has been actively involved in the communities it serves for decades. The company and its predecessors were leaders in providing public access channels and studios in their markets. They also provided schools with vast educational resources through Cable in the Classroom, an industry-wide program.

Educators over the years have viewed Time Warner Cable as an ally in addressing their communities' educational needs. "When I first came to Time Warner Cable in 1993 as a consultant, my job was to coordinate the Cable in the Classroom program on Staten Island. I worked with 100 schools," said Karen La Cava, manager of public affairs in New York City. Over time, her responsibilities expanded to include all of New York City and more than 900 schools. Across the country, tens of thousands of Time Warner employees have joined paid staff like La Cava in philanthropic works, volunteering their time to serve neighbors in their community.

Times change, but the challenges facing the communities Time Warner Cable serves too often remain stubbornly entrenched. That is especially the case in the education field. As company leaders recently reviewed how they could achieve a heightened, sustained impact in terms of outreach, and relate the effort to the company's core businesses, they quickly settled on a glaring, well-known deficiency in American culture: education in science, technology, engineering and math, which is referred to by the acronym STEM.

CONNECT a MILLION MINDS™

AN INITIATIVE OF TIME WARNER CABLE

Chief marketing officer Jeff Hirsch, back center, and director of government relations Marion Boykin, right, observe students exploring STEM studies at a Time Warner Cable-sponsored technology center.

Karen A. La Cava is manager of public affairs in New York City, overseeing Time Warner Cable's Connect a Million Minds (CAMM) initiative and community philanthropy in the New York City area.

"It's well documented there's a big issue there in our country where we've fallen way behind other countries," said Glenn Britt. "It's part of a broader educational problem, but in particular, if we are going to remain competitive with the rest of the world, we have to do a better job in science, math, and technology education, which really related to our businesses and both our products and our employee base. We need employees who have this literacy."[69]

Time Warner Cable decided that the most effective way to devote resources to addressing the issue was to focus on encouraging student interest in STEM education and careers, especially at the critical middle-school age. The company recently created Connect a Million Minds, a multi-faceted program to fund STEM initiatives and connect students to STEM opportunities. The CAMM website, for instance, enables users to type in any zip code in the country to find extracurricular activities in that community relating to STEM education. Time Warner Cable committed $100 million in cash and in-kind efforts to the Connect a Million Minds initiative.[70]

"We have two goals for Connect a Million Minds," said Ellen East, executive vice president and chief communications officer. "One, we want to help parents understand the critical importance of their children's math and science education to a successful future—their own and the country's. And two, we want children to see that math and science can be fun through engaging, hands-on experiences. We hope that will inspire students to pursue higher education and careers in STEM fields, filling a critical but dwindling pipeline in America's workforce."[71]

Experience it. Use your smartphone and watch a video of Glenn Britt describing what it was like growing up as a member of the *Sputnik* generation and the importance of math and science education.

CNN's phenomenal success by the mid-1990s made it the undisputed leader in cable news programming.

Chapter 5

Transitions

1995-2000

Time Warner CEO Jerry Levin ignored Wall Street advice and dramatically increased the company's investments in cable distribution as well as programming in the mid-1990s, positioning the company for a period of explosive growth by decade's end.

Jerry Levin went for a walk in the woods near his Vermont retreat one afternoon in the summer of 1995. He realized that if some on Wall Street knew what he was doing they would have assumed he had lost his way—figuratively if not literally. Well into his third year as chairman and CEO of Time Warner, he was criticized for not producing more value for shareholders. The 1992 Cable Act remained a wet towel damping investor enthusiasm in the cable business. Time Warner Cable executives were making significant progress in getting regulators to take a more even-handed approach to the business, but it would be another year before legislation was passed that opened up new avenues of growth for cable.

Sell some assets, Levin was advised. What about your 19 percent stake in Turner Broadcasting? Buyers, including Rupert Murdoch's News Corp. and GE, were believed to be circling Turner, looking for an opportunity to pounce.[1] And Microsoft's Bill Gates was talking with Turner about an Internet deal involving CNN's web site.[2]

"This doesn't make any sense," Levin said to himself as he wound his way back to his home. He had wanted to help rescue Turner nearly a decade earlier precisely because he hadn't wanted CNN, Turner's crown jewel, to fall into what he considered journalistically unfriendly hands. Plus, he had to admit, he hadn't wanted someone else to own Turner outright if Time Warner couldn't, either.[3]

Levin agreed with his critics that a game-changing move was needed to jump-start Time Warner's fortunes. But instead of selling the stake in Turner, he decided to double down on cable programming and buy the entire company. Also contrary to conventional wisdom on Wall Street, he increased Time Warner Cable's distribution capacity with two significant purchases in a matter of months.

These were followed by sizeable additions over the next few years. From a technology and innovation angle, Time Warner Cable became an industry leader on the path to the broadband future with the debut of Road Runner® and the transition to digital transmission. Transitioning from a single-product to a multi-product company, Time Warner Cable vied for bragging rights as one of the nation's largest and most innovative cable providers by decade's end, even as the major transformational event in corporate history loomed on the horizon.

RIGHT
Jerry Levin, right, and Ted Turner clasp hands as they announce the 1995 agreement for Time Warner to buy the 81 percent of Turner Broadcasting it didn't already own for $6.2 billion in stock, making Turner Time Warner's largest individual shareholder.

BOTTOM
John Malone, left, who held major stakes in TCI and Liberty Media, agreed in 1993 with Bell Atlantic CEO Ray Smith to sell the cable distribution and programming companies to Bell Atlantic. The deal fell through the following year, as Bell Atlantic realized the full cost of upgrading TCI's outmoded systems.

"Don't Mess With My Husband!"

Levin was well aware that Turner perceived him as stifling numerous initiatives Turner had proposed to expand the business, including buying various broadcasters, since 1986 when Time Warner joined the effort to rescue Turner Broadcasting. In an effort to reestablish a personal rapport with Turner, Levin flew out to Turner's Flying D Ranch in Montana by himself in mid-August 1995 to broach the idea of Time Warner buying Turner Broadcasting. Time Warner president Dick Parsons and chief financial officer Richard Bressler remained in New York to work out the fine points of the deal.

Turner's wife, Jane Fonda, arrived unexpectedly at the airport to pick up Levin. Since he had told them he was coming but not the purpose of the visit, she was more than a little guarded in her welcome. "Don't mess with my husband," she warned the surprised CEO. "No, no, no, this is going to be a good meeting—not to worry!" Levin assured her.[4]

Levin tried to meet confidentially with Turner while the gregarious host was entertaining several guests, including NBC network news anchor Tom Brokaw and his wife and daughter. Levin made the case for joining forces to create the country's most powerful vertically integrated cable company. He explained to Turner that it would be a stock for stock deal—valued at $6.2 billion, with Time Warner assuming nearly $2.8 billion in Turner Broadcasting debt as well—and that Turner himself would be the largest individual shareholder, owning 9 percent of Time Warner's stock. He would also be vice chairman.[5]

Intrigued by the prospect of adding value to Time Warner, and exhausted after years of trying to beat Levin at his own game, Turner decided it was time to join him.[6] Levin struck a separate deal to exchange Time Warner shares for John Malone's Liberty Media stake in Turner Broadcasting. (To allay government anti-trust concerns, these were non-voting shares.)[7] The proposed purchase of Turner Broadcasting was announced the following month.[8]

Experience it. Use your smartphone to watch a video of Nick Nicholas describing Ted Turner's legendary role as a cable programming pioneer.

Time Warner's products and services continued to expand as the 1990s progressed, as shown in the 1998 annual report.

Consumer demand for cable news programming, coupled with Bill Gates' growing interest in cable's broadband capabilities, led to the creation of MSNBC in 1996, a joint venture between Gates' Microsoft and the NBC network.

Ted Turner, left, engaged in a war of words with rival Rupert Murdoch, right, founder of Fox News, which became CNN's most formidable competitor. In the middle was Robert Wright, chairman and CEO of NBC, whose MSNBC joint venture competed with both CNN and Fox News.

"You Guys String the Wires and We'll Make 'Em Sing."

Time Warner's purchase of Turner Broadcasting underscored the marquee role played by programming in the cable industry by the mid-1990s. Driven in large part by the availability of HBO and TBS shortly thereafter, the move into urban markets helped differentiate cable from broadcast television. Turner understood the symbiotic relationship between cable operators and providers of original programming earlier than most. "I always used to say to the cable operators ... You guys string the wires and we'll make 'em sing."[9]

The capacity of cable systems continued to expand, and with even more channel space available as digital cable migrated out of the lab and into the living room, programming was much in demand. The '92 Cable Act required cable companies to make programming they controlled available to satellite providers as well. By the late 1990s, more than 100 major cable channels had been created to fill the void.[10] And Turner Broadcasting, with its programming ranging from 24-hour news to cartoons, sports, and classic movies, was the cable programming brand par excellence. Leveraging the value of such a vertically integrated cable company challenged Time Warner and the industry for years to come.

First-Amendment Milestone

In gaining Federal Trade Commission approval in 1996 for the purchase of Turner Broadcasting, Time Warner agreed that Time Warner Cable would carry a cable news channel that competed with Turner's CNN on at least half of its systems. Although Time Warner Cable would have carried additional news channels in any event, the two main contenders for the next launch were MSNBC (a joint venture created by Microsoft and NBC in the wake of Turner's decision to merge with Time Warner) and Fox News, which was being launched in 1996 by News Corp. In September 1996, Time Warner Cable announced that it would carry MSNBC. It said it would add only one news channel at that time, but didn't rule out adding Fox News or another news channel at a future date.[11]

Fox News cried foul. Company officials alleged in interviews, and later in court, that Turner, who had a long-running grudge against News Corp.'s Rupert Murdoch, specifically nixed putting Fox on Time Warner Cable's largest system in New York City so as to keep the competitor from gaining much-needed visibility with opinion makers and advertisers. Turner maintained that he had played no role in the programming decision. Murdoch's *New York Post* started an aggressive anti-Turner campaign, at one point depicting him in a straitjacket and questioning his sanity. Turner took the bait, publicly challenging Murdoch to fight it out in the boxing ring, among other retorts.[12]

The issue transcended name-calling to give rise to important First Amendment judicial rulings for Time Warner Cable, and the cable industry, when New York City Mayor Rudolph Giuliani weighed in on the side of Fox News. With News Corp. threatening to pull more than 1,400 jobs out of New York City, the mayor decreed that Fox News would be carried on one of Time Warner Cable's government access channels, which were designed to carry non-commercial content.

LEFT
New York City Mayor Rudolph Giuliani, right, speaking with Disney Co. Chairman and CEO Michael Eisner.

BOTTOM
Marc Lawrence-Apflebaum, then litigation head for Time Warner Cable, and current general counsel, played a lead role in helping make the cable company's First Amendment case against New York City and Mayor Rudolph Giuliani.

"We Can't Let Them Do This"

When Levin learned of the decision, he was outraged. "We can't let them do this," he told a rapid response legal team he had assembled, including Time Warner Cable CEO Joe Collins, the cable group's then-litigation head (and now general counsel) Marc Lawrence-Apfelbaum, Time Warner general counsel Peter Haje (currently Time Warner Cable's lead director), and outside counsel Robert Joffe of Cravath, Swaine & Moore. The group moved to Cravath's Midtown offices a few blocks away from Time Warner headquarters and spent all night crafting arguments for a temporary restraining order against Fox News and the city.[13]

Joffe led Time Warner Cable's spirited argument, describing Mayor Giuliani's actions as a clear attempt by government to curtail the First Amendment rights of citizens. Judge Denise Cote listened intently to arguments from both sides, then told them she would rule on the matter shortly. Her decision granting a temporary restraining order was a clear victory for Time Warner Cable. As the *New York Law Journal* reported, the judge "concluded the Mayor had deprived Time Warner Cable of its editorial discretion, which is protected by the First Amendment."[14] The decision was ultimately affirmed by the Appellate Court.[15] When Joffe sent Lawrence-Apfelbaum the customary bound copy of the proceeding and related documents a few years later, he included a personal note: "This is why we all went to law school."[16]

Time Warner Cable won an important legal case on First Amendment grounds in 1996 as it argued successfully that New York City Mayor Rudolph Giuliani could not force it to carry Fox News on one of the government access channels of its New York City cable system. Fox News was later added as part of normal system expansion.

Buyer's Market

While the purchase of Turner Broadcasting dramatically increased Time Warner's programming capability, Levin was also actively pursuing deals to greatly expand the company's cable distribution footprint and subscriber base. Two longtime business partners and highly respected cable system operators were at the top of his list.

Advance/Newhouse

HBO had a tradition of hosting annual Super Bowl parties for the cable operators carrying its network. At the 1994 party, Bob Miron, CEO of Advance/Newhouse Communications, Inc., struck up a conversation with Levin. In light of the deteriorating financing environment for cable following passage of the '92 Act, Miron was concerned that the Advance/Newhouse systems, with about 1.2 million subscribers spread across 17 states, needed to be consolidated to gain economies of scale. Finding a partner looked like the best solution. And Levin and Time Warner Cable looked like the best partner.

"(Levin) talked about his dreams for Time Warner Cable. They matched a lot of what my thoughts were," Miron said. "Time Warner Cable had systems I knew in Florida around us, they had systems in New York State, they had systems in the Carolinas, they had systems in Texas, so I knew that we had some pretty good synergies at our locations," he added.[17]

"Barry... Do You Want to Be Known as the Man Who Disappointed 1,200,000 Women?"

As Time Warner Cable's New York City system was upgraded as part of the company-wide capacity expansion beginning in the mid-1990s, 12 additional channels were scheduled to be available for viewing. By the spring of 1997, the competition among fledgling cable channels to get on the system in the nation's media capitol was intense. Going live on Time Warner Cable's system in New York was especially important to programmers because it would provide additional exposure to advertisers and critics. For Time Warner Cable, the 12-channel package could mean additional customers and an additional $3 to $7 a month in subscriber fees, on top of the typical $35- to $40-a-month bill.

Barry Rosenblum, left, president of Time Warner Cable's New York City system, was constantly courted by companies wanting to have their programming carried in the nation's media capital. Pictured at right is technician Jose Davila.

New York City cable system president Barry Rosenblum bore the brunt of the lobbying campaigns. One of the most memorable was from the Romance Classics channel, which carried romantic movies, TV miniseries, and specials. On the day channel executives made their formal pitch to Time Warner Cable, they delivered a truckload of flowers to Rosenblum's office. "Anybody who had allergies sniffed all day," Rosenblum told the *Wall Street Journal*.[18]

The flowers were followed by full-page ads in the New York press directed at Rosenblum and asking, "Barry... Do you want to be known as the man who disappointed 1,200,000 women?" The ad encouraged women to "Call Barry Now," listed an 800 number for the company in New York City, and told them to say, "Barry, I want romance."[19] Despite the full-court wooing, Time Warner Cable spurned the channel's advances and chose a rival suitor.

Venturing Abroad

With the '92 Act curtailing projected returns on cable investment in the United States, Time Warner Cable and other operators started looking abroad for investment opportunities. The timing was good. The strategy, while in tune with the trend toward globalization that was all the rage, needed some work, however.

Many European and Asian countries were stepping up cable distribution of television signals as the 1990s unfolded and were for the most part open to foreign investment to help build out the required infrastructure. Within a few years, Time Warner Cable had investments in France, Hungary (where Home Box Office was also a partner), Sweden, New Zealand, and Japan. Some were successful investments, some less so. The opportunistic investments were never fully integrated into a broader business strategy, however, and were gradually unwound or sold off over time. By the end of 1999, Time Warner Cable no longer owned or held an interest in any cable systems outside of the United States.[20]

The companies continued talks throughout 1994 and into 1995, agreeing on April 1, 1995, to form the Time Warner Entertainment-Advance/Newhouse partnership, referred to as TWE-A/N, to operate systems serving nearly 4.5 million subscribers, including all of the Advance/Newhouse systems. Advance/Newhouse held a one-third interest in the partnership, and Time Warner Entertainment held a two-thirds interest and served as the managing partner.[21] With the exception of Miron and a few other senior executives, all of Advance/Newhouse's employees became employees of the partnership. "With that partnership, we got the scale, we got the geography that we wanted, we got the engineering," he said.[22]

Time Warner Cable was clearly the senior partner in the agreement, yet treated Advance/Newhouse as an equal. The close working relationship that followed was a testament to the wisdom of the initial deal. The companies negotiated terms that provided for an equitable distribution of new systems that Time Warner Cable might acquire to the partnership. "Any new system that came into the Time Warner fold—if it was in the DMA [designated market area], the area surrounding our cable systems—would go into our partnership. If it was in the DMA of a system that Time Warner owned and operated outside of the partnership, it would go in theirs. If it was in neither, we would divide it proportionately," Miron noted.[23] The partners also structured an equitable separation agreement, which came into play in the following decade.

Alan Gerry's CVI

Like Bob Miron, Alan Gerry, founder and sole owner of CVI, had worked with Time Warner and Time Warner Cable executives for years, including Levin, Collins, and Britt, and had a great deal of respect for the management team. The feeling was mutual. Indeed, in three geographic areas where Time Warner Cable and CVI operated neighboring systems—Syracuse, New York; the Raleigh-Durham, North Carolina, Research Triangle; and the Orlando, Florida, area—CVI had fiber-optic systems in place in the suburbs that had greater capacity than the nearby urban Time Warner Cable systems.[24]

As the mid-'90s approached, so did Gerry's 65th birthday. He was one of the hardest-working operators in cable and had taken only a modest amount of cash out of CVI over the years. This could be a good time to get off the horse and try to do some other things, he decided.[25]

Gerry started talking with Time Warner Cable in 1994 about forming a partnership, but the company convinced him that CVI had greater value if Time Warner could buy the entire company, which served 1.3 million basic cable subscribers. They closed on the sale of his business in January 1996. "They had created a special dividend-paying stock for us, so while we had the stock, we would have the growth of the stock and have the dividends," Gerry noted. "It was a whole lot more than what I was taking out of the company," prior to the sale, he added.[26]

Alan Gerry, left, pictured with cable industry pioneer Robert Tarlton in 1997, sold CVI, his cable company headquartered in upstate New York, to Time Warner in 1996.

Gail MacKinnon, executive vice president and chief government relations officer, moved from Time Warner to Time Warner Cable in 2008 as part of the company's effort to ensure that it speaks with one voice to legislators and regulators.

Known for treating valued employees well, Gerry shared the wealth. He distributed nearly $100 million to "many of our loyal people who had been with us. We had a great track record for putting together good people and keeping them," Gerry said. "A lot of our people went off and became regional VPs with Time Warner. It was a great career move for a lot of the younger guys," he added.[27]

During the period Time Warner was talking with Miron and Gerry, it also restructured the ownership of the TWE partnership. In the fall of 1995, Time Warner engaged in transactions in which it bought out the TWE interests of its Japanese partners Itochu and Toshiba. As a result, Time Warner controlled 74.49 percent of the equity interests in TWE, and US West controlled 25.51 percent.[28]

The 1996 Telecommunications Act

While Time Warner executives were on an acquisition spree in the mid-1990s, company and industry lobbyists were spending much of their time in Washington, D.C. Under pressure from the local phone companies as well as AT&T, who wanted to get into each other's businesses, Congress was readying what in many ways was the most sweeping rewrite of federal telecommunications law since the 1930s. The cable industry wanted to do as much as it could to influence the final legislation as it related to their business.

Time Warner Cable, having established its industry leadership position in Washington, D.C., with the Social Contract and as part of the industry's on-time guarantee, led the deregulatory charge to roll back the most onerous aspects of the '92 Act. Gail MacKinnon, Time Warner Cable executive vice president and chief government relations officer, was a lobbyist for Turner Broadcasting at the time of the debate leading up to the '96 Act. It was clear to MacKinnon that "Time Warner Cable drove the agenda at the NCTA" during this period.[29]

The 1996 Telecommunications Act, signed into law by President Bill Clinton on February 8, 1996, reversed many of the more onerous restrictions placed on the cable industry by Congress in 1992, though retransmission consent and must-carry provisions of the earlier legislation remained in place.

Decker Anstrom, left, was promoted to president of the National Cable Television Association in 1993 and worked with industry leaders, including Bob Miron, CEO of Advance/Newhouse Communications, right, to take a more cooperative approach to regulators on Capitol Hill, which paved the way for the passage of the 1996 Telecommunications Act.

The cable industry was "not necessarily the driver behind the legislation," MacKinnon allowed. "But once there's any rewrite of the Telecom Act—they don't happen very often—there is the opportunity to get in there and to try to address the issues that concern you."[30]

The NCTA's second in command, Decker Anstrom, assumed the presidency of the group in 1993 with a mandate to improve cable's image on Capitol Hill and with regulators. Anstrom convinced the group's more recalcitrant, smaller members that they had to be willing to embrace competition if they wanted to be able to push back on some of the more onerous provisions of the '92 Act. If the industry wanted to offer telephone and other services over cable, it had to be willing to let the telephone companies into the cable business.[31]

The November 1994 elections put Republicans in control of both the House and Senate for the first time in 50 years. Deregulation was in the air.

Through much of 1995, telecommunications bills worked their way through both houses of Congress. By June, the Senate had passed its version. The House followed suit in August. The Clinton White House threatened to veto the legislation unless it addressed the administration's concerns about certain aspects of rate regulation and concentration of ownership. The administration also pushed for inclusion of Senate provisions requiring the wiring of schoolrooms and libraries, which was not included in the House version.

Cable Victory

The final version of the Act that President Clinton signed into law on February 8, 1996, was, on balance, a clear victory for the cable industry. Most rate regulation was to be phased out by March 31, 1999. Systems that could demonstrate they faced effective competition, except from satellite, could request to have price controls removed before that date. The legislation also

Netscape Communications helped make the Internet part of everyday life around the world with the introduction of its Navigator web browser in 1995.

freed cable companies to offer phone service, while phone companies received a green light to offer video services. The cable industry once again shifted into growth mode.

Other aspects of the law acted as a drag on the industry. Programmers vertically integrated with cable operators (but not with broadcast networks or satellite distributors) continued to be subject to program-access rules, even though the need for the rules was questionable, as was their application only to cable-affiliated programmers. The industry continued to be required to provide the programming it developed and owned to satellite competitors. And the controversial retransmission consent and must-carry provisions of the '92 Act were retained.

In the first several years in which the law was in effect, retransmission consent generally resulted in deals in which cable operators agreed to carry additional cable networks in exchange for these rights. This barter system led to the launch of many new cable channels, including ESPN's creation of ESPN Classic, and Fox's creation of FX. But in more recent years, retransmission consent has sparked extensive demands on the part of broadcast providers for cash payments, and negotiations which sometimes result in highly publicized "blackouts" or threats of blackouts in the midst of retransmission consent disputes.[32]

"What's the Internet?"

While company representatives and the NCTA pushed for regulatory relief in Washington that everyone hoped would trigger a new growth cycle for cable, Time Warner Cable executives were keeping a tight rein on costs. Not only was the future uncertain, they were still paying for the gilt-edged FSN Jerry Levin had unveiled with such fanfare in 1994. With that in mind, Glenn Britt and Carl Rossetti could be forgiven for keeping their hands on their wallets as they met with representatives from Intel Corp. in 1995. As president and CEO of Time Warner Cable Ventures since 1993, Britt was charged with overseeing TWC's high-tech initiatives. He delegated much of the near-term oversight to SVP Rossetti. Both were feeling the heat from the over-budget and over-hyped FSN. The next big thing was going to have to be pretty spectacular to get them to commit more than a token amount of time and resources.

As far as the visitors from Intel were concerned, the FSN wasn't the last word in cable technology. It was just the beginning. "You guys could offer high-speed access to the Internet," one of the visitors from Intel pointed out to Britt and Rossetti. "We literally said, 'What's the Internet?'" Britt recalled. "We had no idea, never heard of it."[33]

Rossetti took on the task of learning about this latest technology, which had only expanded beyond its academic and research roots beginning in 1993 with the launch of the Mosaic browser and Netscape. "Carl was keeping me up to date on all of this, and it was a little bit off to the side because we had this big FSN thing," Britt recalled. "So my instructions to Carl were, 'Pursue this, but don't spend too much money.'"[34]

Road Runner

Britt and Rossetti decided that broadband "might be a real thing" and started to look for a location to build a trial installation. Stung by the excessive publicity surrounding the delays in getting the FSN up and running, they decided to find a lower-profile launch pad than Orlando, Florida. Elmira, New York, "a little off the beaten track," as Britt said, but with a recently upgraded hybrid fiber coax network , was chosen as the test site.[35]

The Elmira broadband test in 1995, known initially as "The Southern Tier On-Line Community," and later renamed Line Runner, was a success. (Later, as the service was rolled out division by division, it was rechristened Road Runner.) Having learned their lesson in Orlando, Time Warner Cable executives kept the news of the test low-key. It was announced at a glitz-free event at a country club outside Elmira. "Unlike the Orlando FSN launch the year before, almost no press attended," recalled Mike Luftman, who was vice president of corporate communications for Time Warner Cable at the time.[36]

The broadband test "just blew the doors off dial-up modems" in terms of speeding users to the Internet, said chief technology officer Jim Chiddix. "We were already rolling out hybrid fiber coax upgrades at a good clip. And where we had HFC in place, we could roll out the Road Runner service," he said.[37]

Time Warner Cable wasn't alone in its growing interest in the Internet. Several months after the Elmira test, a consortium of cable operators TCI, Comcast, Cox, Cablevision, and others launched an Internet service provider, or ISP, called @Home®.

Synergy by Any Other Name

Time Warner executives were challenged during the 1990s to realize the synergies between their businesses that had been confidently projected by supporters of the 1990 merger of Time Inc. and Warner. One high-profile, unexpected example of corporate synergy came from the merging of brand interests between the cable business and Warner Bros.' Looney Tunes cartoon characters.

When Time Warner Cable executives quickly settled on Road Runner as their top choice for branding the new service that would be rolled out nationwide beginning in 1996, Time Warner CEO Jerry Levin and president Dick Parsons loved it as well. All that was left to do was for a contingent of cable executives to fly out to Los Angeles, meet with their Warner Bros. counterparts, and strike an intra-company licensing deal for use of the Looney Tunes cartoon character's name and likeness.

Britt and Rossetti led the cable delegation. Huddling with Joe Collins before they left, they had agreed that they could go to 10 cents a subscriber per month to clinch the rights to the Road Runner brand. Multiply that times millions of customers a few years out and you're talking real money, they reasoned. Warner Bros. executives had a different idea. The nine cartoon characters they owned generated sizeable revenues per year. That meant the cable team should pony up a significant amount of cash to license Road Runner, many times what the cable team had projected as their top-end bid.[38]

The cable team returned to the East Coast in a state of shock. Dick Parsons stepped in to settle the matter; enough was enough. "I came from a different environment, and while I'm somewhat diplomatic and I love to cajole, and I have a certain amount of patience, some things just have to be," he said. "That was an internal transfer model that just was not going to work, that was just a bridge too far."[39]

While old habits remained slow to change in some quarters, the Road Runner decision under Parsons' leadership, plus a change in management at Warner Bros., helped create a closer working relationship between the traditional Time and Warner businesses going forward.[40]

Sketching the Future

A schematic sketch drawn on a napkin by Time Warner Cable technologist Mario Vecchi at a January 1995 dinner meeting helped create the architecture for what became the Road Runner broadband service.

Carl Rossetti called together a small group of Time Warner Cable technology experts on January 25, 1995, to hash out a plan for getting the company into the Internet access business. Over dinner at Jim Chiddix's mountain home in Evergreen, Colorado, he and Chiddix discussed various technological options with Vecchi, who had worked with Mike Hayashi on the digital modems used in the FSN and took the lead in developing Time Warner Cable's Internet access technology, and Tim Evard, who would quarterback the rollout of the business.

The result of their dinner meeting was a sketch by Vecchi of how such an Internet service would work. Months of work followed as the design was amended and refined, but the sketch remains an iconic image of Time Warner Cable's innovative technological prowess at work.[41]

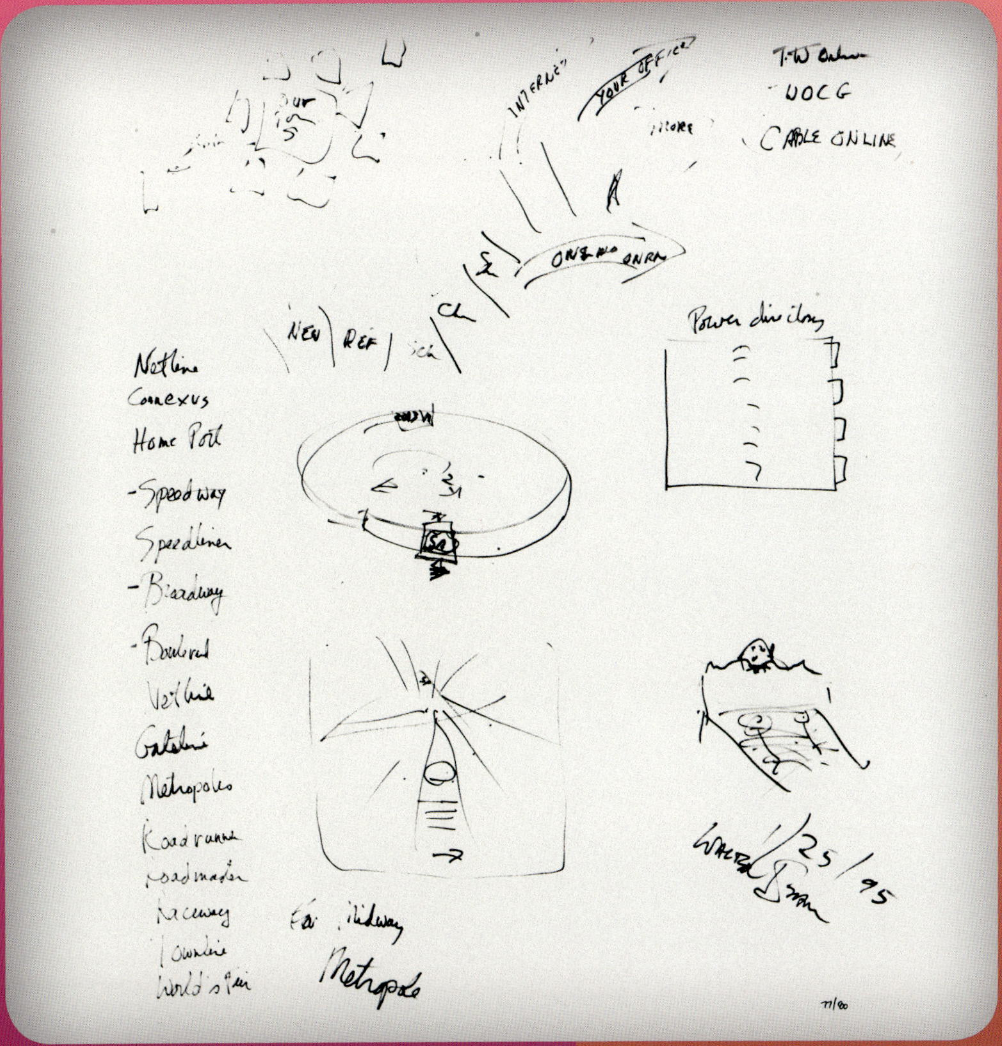

TOP
A schematic sketch drawn on a napkin by Time Warner Cable technologist Mario Vecchi at a January 1995 dinner meeting helped create the architecture for what would become the Road Runner service.

LEFT
Then-senior vice president Carl Rossetti championed Time Warner Cable's development of broadband Internet technology, leading to the creation of its Road Runner service. Rossetti is now president of Time Warner Cable Ventures.

Road Runner, the first cable-delivered high-speed Internet service, launched in Akron and Canton, Ohio, in September 1996. It would soon spread into additional areas throughout the country. In New York City, above, the launch included a Road Runner–costumed cartoon character greeting Time Warner Cable's Manhattan officials, including from the right, Howard Szarfarc, Roosevelt Mikhail, and Cesar Beltran.

Road Runner Takes Off

Rising to meet unexpected challenges has long been a part of the Time Warner Cable culture, and it once again exhibited itself as Road Runner got up to speed. Steve Fry, president of the Northeast Ohio division, received a call one day in mid-1996 from Time Warner Cable president Jimmy Doolittle. Since the Elmira test, Time Warner Cable had been pulling together a team of hardware and software providers—with Hewlett-Packard developing routers and servers, Motorola the cable modems, and Microsoft its Internet Explorer browser—to create a more robust Road Runner product.

It was time to officially launch Road Runner. "We want to be able to bring an entire division up at one time, not roll it out neighborhood by neighborhood. Let's create a product that people can buy like they buy HBO," Doolittle said. He wanted to know if Fry wanted his division to be the first to take on the challenge. Fry did.[42]

Augmented with employees from the equipment and software contractors, Fry gave his team barely two months to move from project concept to reality, and gave them the authority to make it happen. "With projects like this, I would drive a stake in the ground and say, 'This is when it's going to be done, and I know some things have to happen in between. Go ahead and work on it.'"[43]

The team came through, and on September 10, 1996, Fry and Time Warner Cable CEO Joe Collins unveiled Road Runner, the first cable-delivered high-speed Internet access service. "We launched it across all of Akron and Canton on the same day. Mission accomplished, we had it up and running, and it worked. People could call in and get it," Fry said.[44]

Each of the Time Warner Cable systems that launched Road Runner came up with its own solutions, even though it was essentially the same high-speed service wherever it was launched. In Columbus, Ohio, division president Terry O'Connell poached a well-known Internet expert from Internet service provider CompuServe, also based in Columbus, to help his team get the new service up and running in 1996. "We had lots of help from the corporate folks, but I will tell you, in the field we launched it 30 different ways," O'Connell said. "It was just an enormous effort that involved every department in the business," he added.[45]

"Something's Backward Here"

For at least the first year of rolling out Road Runner, Time Warner Cable did not fully appreciate the extent to which customers were increasingly interested in using the service to connect to the Internet, rather than to get access to content produced or aggregated by Time Warner. The launch in Akron and Canton, for instance, made much of the various Time Warner news and entertainment sites the customer could connect to. Time Warner Cable had also lined up local news providers in each market as the service was rolled out.

As interim CEO of Road Runner, Carl Rossetti was reviewing expenses and "one of the first things I noticed was our expense for connectivity was going through the roof. So I went back to the engineers, and I said, 'What's going on? What's happening here? I mean, we're not even close to our budget for connectivity.'"

Experience it. Use your smartphone to watch a video of Carl Rossetti describing his realization that customers were using Road Runner mainly to access the Internet rather than Time Warner content sites.

His engineers reported back that they had figured out the problem: So many of the customers were using the Internet, which was driving up connection costs. Rossetti said, "Okay, well, we've got to do something about that." The engineers' solution was to put in a "blocker" to stop subscribers from getting to the Internet. "Wait a second," Rossetti said, "something's backward here. Every one of our customers wants to go to the Internet. I don't think the answer is to stop them."[46]

It was a lesson learned by every Internet service provider in short order and helped to propel billions in infrastructure spending to boost access and capacity in subsequent years. And it was also a costly lesson for content providers, including Time Warner, which in a few years shelved its multi-million-dollar Pathfinder Internet site, which was designed to serve as a one-stop gateway to content developed primarily by Time Warner's magazine division.

The Road Not Taken

By 1997, there was no question that Britt and Rossetti had a hit on their hands with Road Runner, even if it was still being rolled out. The only real question was, how were they going to fund the growth of this new service? Britt had already been forced to scrape together funding from numerous internal sources and budgets to get the service up and running. They needed to write some sizeable checks if Road Runner was going to begin to realize its potential.

Britt and Rossetti may have gotten over letting the FSN curb their passion for new ventures, but not Levin. He was still feeling burned by the cost and negative publicity surrounding the Orlando experiment. And he was preoccupied dealing with Ted Turner and Time Warner's latest cable system acquisitions. He wasn't inclined to fund another high-tech venture that—in its early stages, at any rate—looked like an unlimited call on Time Warner's treasury department. "Jerry Levin said he wouldn't fund it," Britt recalled. "We were trying to persuade him to make this a big thing for Time Warner, to make it a big content vehicle."[47]

It was a lost opportunity, Britt said, that might have altered the company's future. If Levin had funded and embraced Road Runner, he might have succeeded in propelling Time Warner into the digital online future. Instead, a few years later, he committed Time Warner to the largest corporate merger ever in an effort to play catch-up with the rapidly accelerating online world. "I would say that if he'd gotten behind it, AOL would never have happened," Britt speculated.[48]

Possible Road Runner IPO

Britt at least got the go-ahead from Levin and Parsons to pursue funding from outside the company. He had a clear sense that his superiors thought it was a long shot. While initially courting Oracle and Intel in early 1998, at the last minute Britt made a deal with Microsoft and Compaq Computer to fund Road Runner.[49] In June 1998, the two invested a total of $425 million in a Road Runner joint venture with Time Warner, TWE, and TWE-A/N as well as MediaOne, which had been formed out of the cable assets of US West. (The following year it was bought by AT&T.) The investment gave Microsoft and Compaq each a 10 percent stake in the joint venture.[50]

"The Feeling of Satisfaction was Amazing!"

Lisa Lavelle
Lead Central Operations
Support Representative
Rochester, NY

I joined the family at Time Warner Cable in March 2001 as a central ops dispatcher and moved to the lead position in my department a few years ago.

My biggest challenge was learning to change at a fast pace. Here things change by the moment. We take new directions and flow wherever we need to, be it changing systems, system upgrades, or job descriptions. We take it all head on and work as a team to understand the process and make it work for the company. For example, when we merged with Buffalo, it was tough trying to get everyone on the same page and ready to handle calls from all of the Western New York division, but we're making it happen.

I had a customer who had been having problems for a long time with his home account and could never meet our time frames for a service call. He was ready to cancel his service. I worked with him and the lead tech in his area, and we did what it took to make his service function fully. We were able to save a customer, and the feeling of satisfaction was amazing!

I came to Time Warner Cable because I kept complaining about the Road Runner bill, so my husband said, "Then go and work there!" And here I am, almost 10 years later. I feel it's better than other companies because we care about our customers and our community.

With outside funding, Road Runner was structured as a standalone business. Doing so had the added benefit of making it easier to cash in on the mania for all things Internet that was gripping the markets by the late 1990s. When the rival @Home Internet service went public in 1997,[51] and later used its sky-high stock price to buy the Internet portal Excite for $6.7 billion in stock,[52] Time Warner determined to position Road Runner so the unit itself could operate independently as the first step to a possible public offering.

Road Runner raced to keep up with customer demand. At the time of the cash infusion from Microsoft and Compaq in June 1998, Road Runner had just crossed the 50,000 subscriber mark.[53] By the end of the following year it had grown tenfold to roughly 550,000 subscribers, and was the largest cable online service in the country. And in keeping with its independent operating model, more than 250,000 of those were customers of cable companies other than Time Warner Cable.[54]

Billion-Dollar Seal of Approval

By the spring of 1997 it had been more than four years since John Malone had boasted about his system's imminent ability to carry 500 channels, as soon as cable converted to digital. More than two years had passed since the FSN demonstrated that digital cable functioned well in a limited test situation. Cable's long-anticipated digital conversion had yet to happen, and Malone, Time Warner Cable, and the rest of the cable operators were feeling the heat.

The Internet was taking off, as were the share prices of start-up companies based on this new medium. No less than Intel CEO Andy Grove had declared several months earlier that he didn't think the country's cable infrastructure was up to handling the surge of Internet-driven communications. Meantime, rival satellite television services were delivering digital signals.

Under the auspices of the CableLabs industry technology consortium, Malone led a group of cable executives on a goodwill tour of high-tech companies in Silicon Valley, and then up to Redmond, Washington, to meet with Microsoft founder and CEO Bill Gates. Among their goals was to highlight the billions they were spending collectively on fiber and other upgrades to their systems. They also engaged in numerous discussions centering on developing and building affordable digital set-top boxes. The group that sat down to dinner with Gates in April 1997 included Malone; Time Warner Cable's Joe Collins; Brian Roberts of Comcast; Cox Communications' Jim Robbins; and Richard Green, the head of CableLabs.[55]

Gates was impressed with the industry leaders and their commitment to broadband technology. He clearly saw video as central to the growth of the Internet and promoted broadband access as crucial to greater use of video. At one point during the dinner, Brian Roberts, half in jest, suggested that with Gates' cash hoard, he could easily buy 10 percent of each of the companies represented around the table. Plenty of laughter followed, and Malone actually needled Roberts later for appearing to all but beg the country's richest man for a handout.[56]

When Roberts returned to his office in Philadelphia, he had a message from Microsoft's vice president for corporate development, Greg Maffei. Gates wanted to follow up on the idea of investing in a cable company, specifically Comcast, which was considered a well-managed, essentially pure play on cable. More meetings followed, and in June 1997, the companies announced that Microsoft was buying 11 percent of Comcast for $1 billion.[57]

The billion-dollar investment from Gates not only sent Comcast's stock price soaring—all the major cable companies saw their fortunes suddenly rise. The investment from Gates transformed Wall Street's perception of cable from has-been to must-have. Cable companies would never achieve the stratospheric share prices of pure Internet plays, but on the other hand, nor would they collapse so completely when the dot.com bubble burst a few years later. Suddenly investors were willing to acknowledge that the industry's commitment to fiber-optic cable, led by Time Warner Cable, and other investments positioned it squarely as a major player in the increasingly Internet-dominated future of telecommunications.

Satellite Television

The cable industry, dominated by entrepreneurs who competed fiercely with broadcasters and the phone company during its early decades, suffered a serious competitive blind spot in the late 1980s and early 1990s. It forgot to look up. By the time it did, satellite television was firmly established as a competitive threat.

Much as the fledgling cable business had benefited from the FCC's four-year freeze on granting broadcast licenses that ended in 1952, the struggling direct broadcast satellite (DBS) business received a regulatory boost of its own 40 years later. The requirement in the '92 Act that cable operators had to provide cable programming to satellite competitors was viewed by them as nothing short of a godsend for the satellite business.[58]

Setting Standards

With the Internet clearly transforming communications by the mid-1990s, the cable industry rallied to form open, common technological standards for creating digital cable modems that would serve as the consumer gateway to the online world. A common standard was crucial to the industry's ability to achieve economies of scale and drive down modem costs for consumers. It was also vital in ensuring that the cable industry would be part of mainstream communications in the 21st century, and not a technological backwater.

Four leading cable companies—Time Warner Cable, TCI, Cox, and Comcast—created Multimedia Cable Network Systems, MCNS, in 1995 and worked closely with CableLabs and others to define open, interoperable specifications for cable modems.

CableLabs, formed in 1988 as the research and development brain trust for the cable television industry, played a key role in brokering agreements on the cross-licensing of modem technologies and setting common modem standards. By March 1997, a group of vendor authors issued the Data Over Cable Service Interface Specification (DOCSIS), which created the basic specifications for all future data modems, and first-generation modems were tested that summer. CableLabs certified the first DOCSIS modems in the spring of 1999.

Successive iterations of the DOCSIS standard reflect the evolving role of cable as the carrier of data and phone service as well as video signals, and CableLabs' contribution to the industry's phenomenal growth.[59]

In fact, satellite was initially viewed as directing more business to cable programming providers like Time Warner. "Ironically, the first couple of years of DirecTV helped the cable business enormously," said Joe Collins. "They were national advertisers, and so they put out a big national advertising campaign, and of course what you're selling when you're selling that product is the cable signals, and so people would hear about HBO and CNN and all this wonderful stuff, and they'd say, 'Let's get that on cable. We've always wanted to think about that. Let's do it now.' So originally it actually helped our business," he added.[60]

Jim Gray, right, who retired from Time Warner Cable as vice chairman in 1994, served as chairman and CEO of PrimeStar Inc., the satellite television service owned jointly by a group of cable companies, from 1995 to 1997. Dan O'Brien, center, was president and chief operating officer of PrimeStar, and Ken Carroll, left, served as chief financial officer of TCI Satellite Entertainment.

While most cable operators were preoccupied with issues stemming from the '92 Act and lobbied in Washington for changes that were incorporated in the '96 Act, the DBS providers grew steadily during the mid-1990s. EchoStar's Dish Network, and DirecTV, owned by General Motors' Hughes Electronics unit, extended their reach with aggressive consumer marketing campaigns. On Capitol Hill, they trumpeted satellite as the only true competition to cable. It was a view that appeared to be widely shared among regulators and many legislators, much to the cable industry's chagrin.[61]

PrimeStar

John Malone led an effort among cable operators, including Time Warner Cable, Cox, Comcast, Newhouse, Viacom, and Continental, to form their own satellite service. Marketed as PrimeStar beginning in the early 1990s, the service was initially targeted at customers in rural areas where it wasn't economically feasible to reach them with cable. Within a few years, it was the largest satellite television company, reaching 2.3 million subscribers by the end of the decade.[62]

Rupert Murdoch had launched his BskyB satellite television business in the United Kingdom to great success. By 1997, he was eager to use additional satellite capacity to break into the satellite market in the United States as well. He initially signed an agreement with EchoStar to buy 50 percent of the business for $1 billion. Outraged that a major provider of programming was about to compete in the distribution business, some cable operators fought back. Eventually, after disputes between

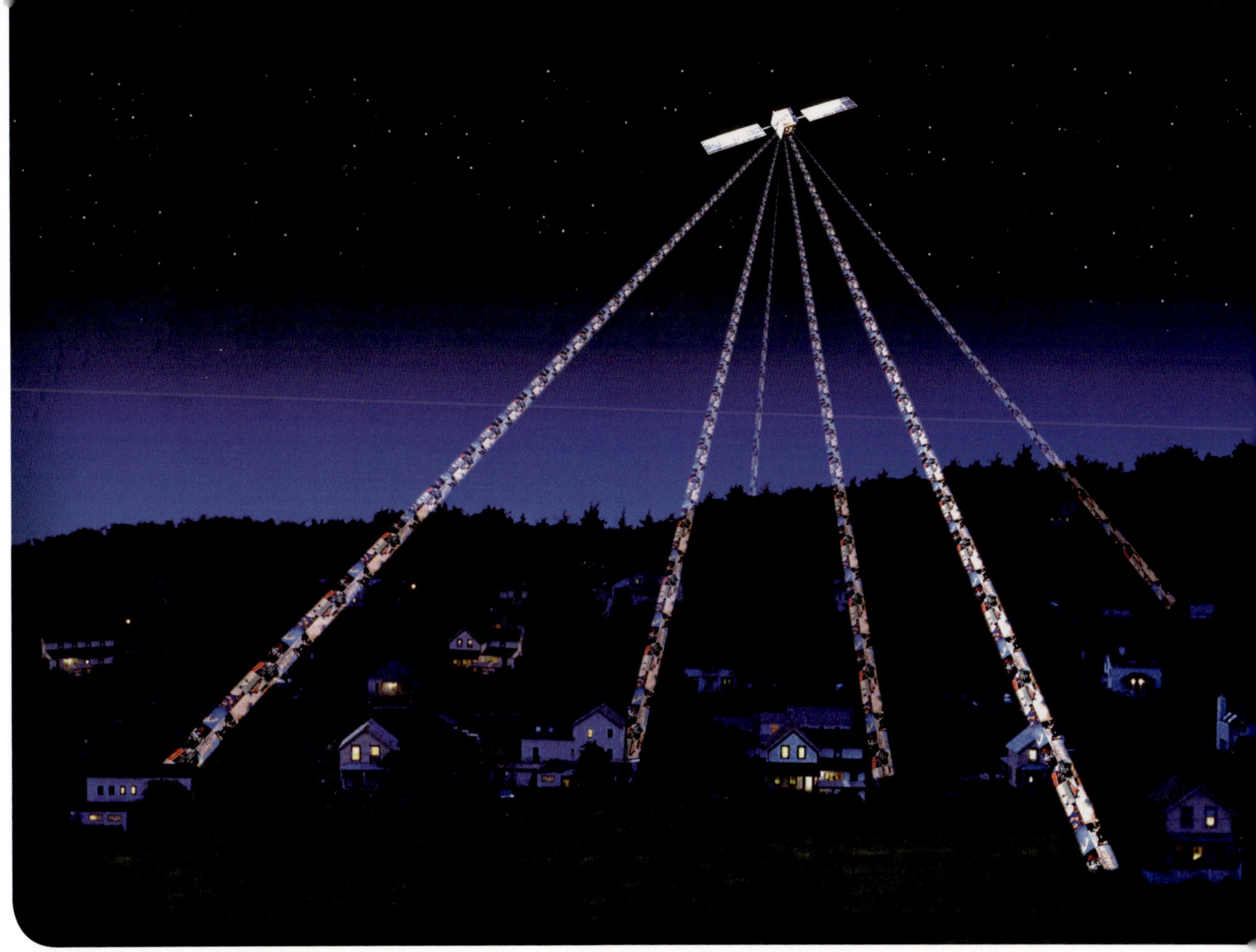

News Corp. and EchoStar over terms of the deal, Murdoch walked away from it. Following that, in June 1997, a satellite joint venture between News Corp and MCI signed a deal with PrimeStar that would have contributed the joint venture's assets to PrimeStar in exchange for a 10 percent stake.[63]

PrimeStar's plans were grounded the following year when the U.S. Justice Department sought to block the PrimeStar—News Corp/MCI deal on antitrust grounds. "DBS presents the first real threat to the cable monopoly," said Joel Klein, assistant attorney general of Justice's antitrust division. "Unless this acquisition is blocked, consumers will be denied competition—lower prices, more innovation, and better services and quality."[64]

Despite the Justice Department's best intentions, its action had the effect of reducing competition among satellite television providers. In January 1999, PrimeStar sold its satellite assets to EchoStar for $1.36 billion. Murdoch, responding in part to a breach of contract suit from EchoStar, sold his satellite assets to EchoStar as well.[65]

Digital Video Service

Having launched and tweaked the FSN in Orlando, Mike Hayashi and his team returned to Time Warner Cable's Denver-area technology lab in the mid-1990s to continue work on the next generation of digital set-top boxes (STBs). The FSN test demonstrated that digital set-top boxes functioned well. The challenge, faced by Time Warner Cable, TCI and all cable operators anticipating converting to digital signals, was to find manufacturers capable of producing boxes at a reasonable price point that consumers would be willing to pay.

Hayashi and his team, working separately from Mario Vecchi and the Road Runner group, had an additional goal. The FSN test demonstrated that two-way boxes worked in real-life conditions, and consumers, at some point, were going to want to have the capability to place real-time orders through their cable service—including ordering video on demand.

PrimeStar provided its cable company owners with an economical way to reach homes that were not part of or adjacent to its increasingly clustered systems around the country.

AthenaTV, a satellite-based joint effort between Time Warner Cable and HBO, provided an efficient, affordable way to receive digital signals from a range of digital programmers as the cable industry was making the transition from analog signals to digital beginning in the late 1990s. Pictured left to right are: Elmer Musser, Home Box Office; Jon Ralph, Home Box Office; Michele Lezama, Home Box Office; Mel Huey, Time Warner Cable; Jim Braun, Time Warner Cable; Dom Serio, Home Box Office; Kevin Leddy, Time Warner Cable; Barbara Jaffe, Home Box Office; John Callahan, Time Warner Cable; John Beyler, Home Box Office; Melinda Witmer, Home Box Office (now executive vice president at Time Warner Cable); and Craig Cuttner, Home Box Office.

Trojan Horse to Pegasus

Set-top boxes the Time Warner Cable team pursued would "have the same characteristics of a one-way device that could offer 500 channels, but be able to wake up and be a two-way device," Hayashi said. He coined the name "Trojan Horse," which conveyed the idea of the hidden capabilities. Chiddix told him it had negative connotations, handed Hayashi a copy of *Greek Mythology for Dummies*, and suggested he try again. Hayashi chose Pegasus. It was still a horse, and the winged feet conveyed a sense of speed.[66]

Digital cable service employing the Pegasus set-top boxes built by Scientific-Atlanta was tested in 1998 and early 1999, and then initially rolled out in Austin, Texas; Tampa, Florida; and Columbus, Ohio.[67] "This was kind of the first genesis of the digital system we use today. For me, the rest is history," Hayashi said.[68]

"We rolled out VOD on top of that by just rolling out servers into the headend," he noted. "By a flick of the switch, you now have video on demand. We were able to offer services comparable to what the FSN was offering just five years [earlier] for a cost at one-tenth to one-twentieth of what it cost us to do it back then. Even then, it was kind of expensive, by the way, but it was manageable."[69]

To provide a cost-efficient means of receiving digital signals from each programmer during this period, Time Warner Cable developed a joint project with HBO called AthenaTV. Transmitting roughly 50 digital video services off of four satellite transponders to Time Warner Cable digital headends, AthenaTV enabled the company to manage costs, plus picture and sound quality, while maintaining control of the programming lineup for customers. The service uplink was located at HBO's facilities at Hauppauge, New York. The project was launched in late 1998 and closed down in mid-2002 once local digital receiving equipment had become more affordable.

"We Have Vision Here"

In 1997, Time Warner Cable added true IP phone service, or Voice over Internet Protocol (VoIP), to its list of technology firsts, launching a test in Portland, Maine. Other operators, namely Cox, had demonstrated there was consumer demand for cable phone service, but they had installed switches like those used by telephone companies to carry calls, rather than using Internet protocol. Time Warner Cable had actually carried out a few such "circuit-based" tests as well using existing telephone technology. The Portland test marked another industry milestone, though cable customers across the country would have to wait a few years for VoIP phone service to roll out nationwide.[70]

There is no question that Time Warner Cable led the cable industry technology pack in the 1990s, said Mike LaJoie, a Los Angeles–based technology consultant to Warner Bros. who joined Time Warner Cable to work on the FSN and succeeded Jim Chiddix as chief technology officer. "We were ahead of the crowd with high-speed data, we were ahead of the crowd with video on demand, we were ahead of the crowd with advanced digital, we were ahead with voice, and were ahead of the crowd with DVRs. We were ahead of the crowd."[71]

Cable's Climb

The 1990s began with cable television networks reaching barely a fifth of American households. By decade's end, cable networks accounted for a greater share of U.S. households than the broadcast networks.

Viewing Share of Households: 1989-2000

(*In millions*)

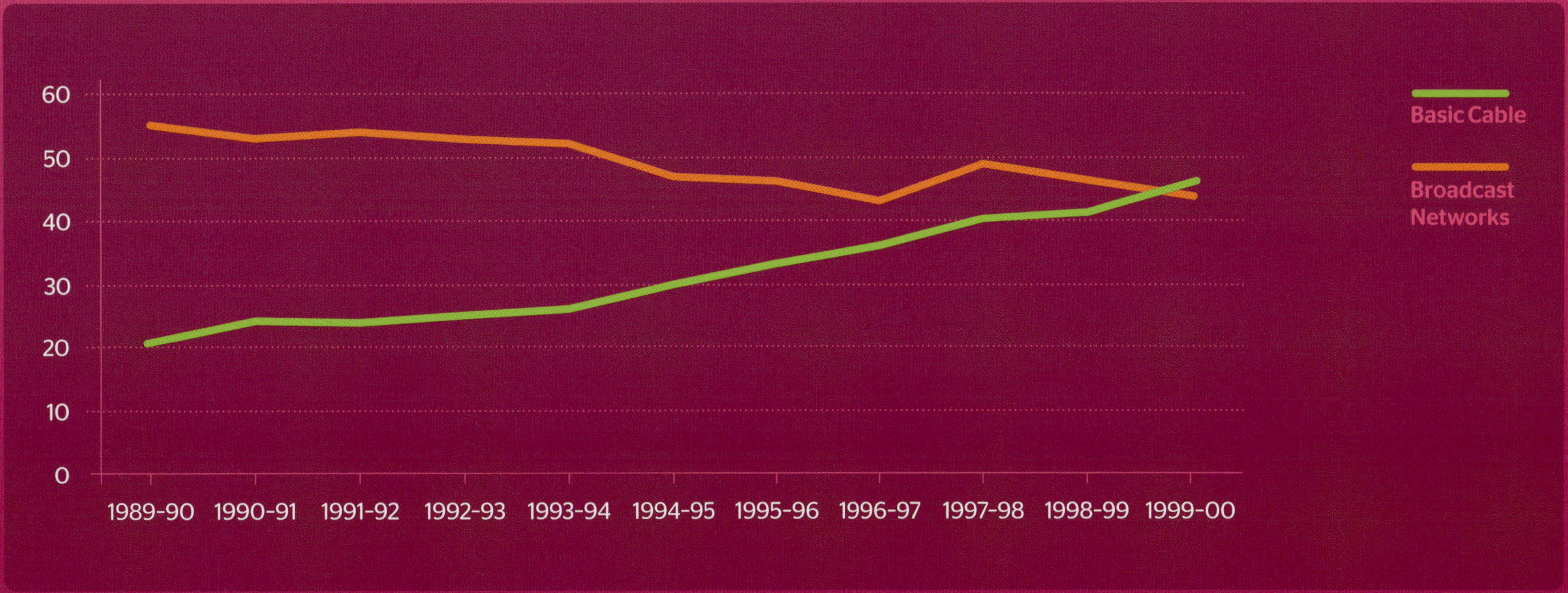

Year	Basic Cable	Broadcast Networks
1989-90	21	55
1990-91	24	53
1991-92	24	54
1992-93	25	53
1993-94	26	52
1994-95	30	47
1995-96	33	46
1996-97	36	43
1997-98*	40	49
1998-99	41	46
1999-2000	46	44

**Broadcast Network share here and following included Fox Broadcasting affiliates, previously counted as independents.*

A. C. Nielsen Co. and the Cable Advertising Bureau; Cable TV Facts. Paul Kagan Associates, Cable TV Financial Databook 2000, p. 35.

Chief technology officer Mike LaJoie said that Time Warner Cable is "still breaking ground and leading the charge" in defining the next generation of broadband consumer experience.

Much of the credit for this blue-ribbon performance, LaJoie said, goes to the talented in-house technologists, including Jim Chiddix, Louis Williamson, Mike Hayashi, Mario Vecchi, and many others. Credit for Time Warner Cable's high-tech prowess also goes to the team of executives leading the cable effort over the past two decades—Joe Collins, Glenn Britt, and Carl Rossetti in particular, said LaJoie:

They were aware that we were at a watershed and that the changes in technological innovation were going to happen so fast that it was going to have a huge impact on their business. They didn't know exactly what the changes were going to be, but they were aware of it. They wanted to be involved in it, and they wanted to be in the forefront of it.

Which actually speaks to the culture of Time Warner Cable. The culture of Time Warner Cable is that we have vision here. We are looking out in front. We are trying to figure out the best way to leverage the combined assets that we have in our investment capacity to bring more and more interesting things to our customers. And that's not something that's new. That's deep in the bones of Time Warner Cable. It's deep in the bones of the cable industry. The cable industry is a very, very innovative industry from way back when, when the first guy said I'm going to put an antenna on a hill and string a wire down that hill to that town with a bunch of amplifiers in it and we're going to get pictures in there where nobody got them before.[72]

Cable operators also pursued the business market for phone service in the late 1990s. TCI bought into Teleport Communications, which offered high-speed optical-fiber links to the long distance network, in 1992. TW Telecom (now called tw telecom inc., a standalone public company) was formed in 1993 as a joint venture with a division of TWE and US West.[73] In July 1998, the business was reorganized as a self-financing entity offering business telephony services, and issued $400 million in debt securities.[74] The following May, the unit sold stock to the public, raising $252 million[75] and reducing Time Warner's stake in the enterprise to about 48 percent.[76]

Glenn Britt played a pivotal role in shepherding the development of many of Time Warner Cable's key technology contributions during the 1990s. His role did not go unnoticed in the Time Warner executive suite. In January 1999, Britt was named president of Time Warner Cable. The promotion positioned Britt as the in-house favorite to succeed Time Warner Cable CEO Joe Collins barely a year before the entire Time Warner corporation underwent the most traumatic transformation in its history.

Let's Make a Deal

A series of mergers, acquisitions, and joint ventures among cable companies in the 1990s left a cat's cradle of interlocking ownership ties as the end of the decade approached. With the economies of scale offered by clustering systems ever more apparent—particularly in terms of the ability to attract advertisers and negotiate with programmers—the need to rationalize system structures was growing more acute by the day. That was particularly true for TCI and Time Warner Cable.

In 1998, Time Warner Cable and TCI, which was being run by president Leo Hindery, a respected cable industry veteran Malone had recruited in 1997 to improve TCI's operating performance, engaged in a series of transactions to rationalize their holdings. In 1997, Hindery had started the process with a number of trades with other operators starting in mid-year, which led the cable TV trade press to dub 1997 the "summer of love."

TCI Chief Executive Officer Leo Hindery, right, recruited by John Malone in 1997 to boost TCI's operating efficiency, engaged in multiple system trades with Time Warner Cable. He looked on in this 1997 photo as TCI Chief Operating Officer Marvin Jones, left foreground, and Drew Fleming, right foreground, TCI of Kansas City general manager, compared "war wounds" from their days as installers. At center rear is Greg Harrison, TCI director of government affairs.

Time Warner Cable's Joe Collins met Hindery for breakfast one morning in 1998 to discuss executing a staggering series of more than 30 trades and joint ventures. "The last part of the trade was we would take all of our Houston properties and put all of them together into a big joint venture," said Collins. "So Leo looks through the list ... reaches in his pocket, takes out his pen, writes his name on the bottom. He says, 'Done,'" added Collins. "By and large," said Collins, "that was the deal we did," following careful review by lawyers on both sides.[77]

As a result of the Houston-area joint venture and another with TCI in Kansas City, both of which were managed by Time Warner Cable, the company added approximately 660,000 subscribers in 1998. That vaulted Time Warner Cable to 12.6 million customers served by its 33 divisions.[78]

Time Warner president Richard Parsons testified on Capitol Hill in 1998 on the possible anti-trust implications of AT&T's proposed $48.3 billion purchase of TCI and Liberty Media.

"Everyone Just Rose to the Occasion"

The flurry of transactions agreed to in 1998 with the stroke of a pen among top Time Warner Cable and TCI executives, plus others that followed in 1999 with MediaOne, Fanch Communications, and Comcast,[79] were made to make Time Warner Cable's systems more cost-effective and efficient. They had that effect. But they also had a lasting impact that reached all the way to frontline employees serving customers, repercussions that might not always have been as apparent.

Carol Hevey recalled the shock she felt as division president of the greater Boston area when she learned in late 1998 that the division was going to be swapped to MediaOne the following year in return for that company's system in Dayton, Ohio. The greater Boston division had posted the highest operating cash flow in the company in 1997 and 1998. "It was a great team of people who worked well together, worked hard every day to build the business, and to take care of customers, and to sell more products. It was everything that you would want a cable operation to be in every respect."[80]

Hevey was worried that when she broke the news of the pending deal to the staff in January 1999 that morale would plummet. But the pride and professionalism of the Time Warner Cable staff came to the fore at all levels. "The deal didn't close until the end of July," she recalled. "So there was the January through July time frame in 1999 when we were all still working together and working to run our business as well as we ever had, knowing that that date was coming. It was an extraordinary experience, actually. Everyone to a person, from the frontline folks all the way to the senior leaders, everyone just rose to the occasion."[81]

Some employees made the switch to MediaOne. Employees also took jobs with Time Warner Cable in one of its other divisions, which totaled 34 by yearend 1999. Still others, including Hevey, left the industry, although in her case, the sabbatical only lasted about 18 months.[82]

Dot.com Boom

Michael Armstrong, a veteran IBM executive who had also headed General Motors' Hughes satellite TV business, took over as CEO of AT&T in the fall of 1997. He was determined that AT&T was going to get on the Internet juggernaut and viewed cable television as the vehicle best suited to make that happen. AT&T had been operating as a long-distance-only business since the breakup of the Bell System in 1984 and needed to make the "last mile" local connection to America's homes.

As the pace of corporate deal making quickened in 1998, and with every deal of any consequence seeming to involve the Internet, Armstrong was anxious not to be left out in the cold. In May 1998, he met with a group of the leading cable industry executives, including Time Warner Cable's Collins, at an industry conference in Atlanta. They discussed working as a group to provide the last mile connection to AT&T. There proved, however, to be a clash of cultures, especially when it came to different approaches to service, between the phone utilities and the cable companies. "No deal was made," Collins said, "but it really sparked an interest on their part in doing this."[83] Armstrong also failed to reach a deal with the cable operators that owned the @Home Internet connection service.[84]

TCI's Hindery courted Armstrong during this period, as did Malone. Hindery in particular was conscious of the fact that the dot.com boom was accounting for much of the appreciation in the stock of TCI as well as most cable television stocks. He wanted to strike while the market was hot. By June 1998 he had succeeded: AT&T announced that it was paying $48.3 billion to buy TCI and Liberty Media.[85]

Fiber-Optic Cable

- The U.S. cable industry's multi-billion-dollar upgrade will bring an array of new and enhanced video, data and voice offerings to consumers. Approximately 40% of U.S. cable subscribers are served by systems with 750 MHz capacity, up from 28% in 1996 and expected to grow to more than 60% by 2002.[1]
- Consumer spending on U.S. cable services is expected to grow 55% over 1997 to $40.3 billion in 2002.[1] Spending on cable modem services is expected to soar from $96 million in 1997 to $5 billion in 2002.[2]

HBO THE WORKS

HBO has increased its commitment to multiplexing to take advantage of cable's expanded digital channel capacity. By May 1999, the network will offer 10 distinctly branded HBO and Cinemax channels.

In 1998 Warner Home Video's DVD wholesale domestic gross sales more than doubled over 1997 and represented more than 10% of those sales for the year.

Digitization

- Following Moore's Law (costs halving every 18 months), the economics of digital technology are becoming compelling. As cable operators are able to store and deliver a vast inventory of movies at reasonable costs, video on demand will become a reality.
- With more than 1.5 million DVD players shipped by manufacturers, the launch of DVD is significantly more successful than that of the CD or VCR at a comparable stage.

"I want it now...

- The Internet is the fastest-growing segment of the media industry, with revenues projected to increase from $9 billion in1997 to $56 billion in 2002.[3]
- The number of online users in the U.S. is expected to grow 84%, from 63 million in 1998 to 116 million by 2002.[3]
- E-commerce revenues are expected to rise from $7 billion in 1998 to more than $41 billion in 2002.[3]

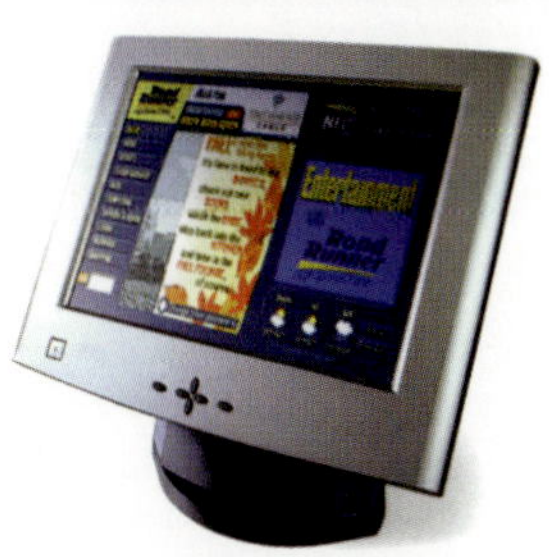

Internet

Road Runner, Time Warner's jointly owned high-speed online service, reached 180,000 customers at year end, delivering data to computers via fiber-optic cable and cable modem at speeds more than a hundred times faster than telephone dial-up services.

TIME WARNER INC. 1998 ANNUAL REPORT 17

This page from Time Warner's 1998 annual report positions the company as a leader in the dot.com world of digitized entertainment and high-speed broadband Internet connections.

America Online Chairman and CEO Steve Case spoke at a press conference on January 10, 2000, announcing AOL's proposed combination with Time Warner. To Case's left are Gerald Levin, chairman and CEO of Time Warner; Ted Turner, vice chairman of Time Warner; Robert Pittman, president and COO of AOL; Richard Parsons, president of Time Warner.

Time Warner and AOL

Jerry Levin was as anxious as Armstrong to make a transformative deal that would catapult Time Warner into the leading ranks of Internet digerati. He didn't think the company was capable of getting there on its own. His thought process echoed the approach he had taken in the months leading up to the Time-Warner merger of a decade earlier, only the technology in question had evolved from movies and video to the Internet. "I just concluded that you couldn't digitize the company, and therefore we needed to inject the company with this Internet perspective," he said.[86]

After initial conversations with Yahoo! founder Jerry Yang went nowhere, Levin met AOL's chairman and CEO Steve Case at a Washington meeting of business executives convened by Treasury Secretary Robert Rubin in 1999. Levin was impressed with the respect Case garnered in Washington and on Wall Street, where AOL stock had soared. That fall, Levin and Case met in Paris, and again in Beijing at a meeting sponsored by Time Warner's *Fortune* magazine, which coincided with the 50th anniversary of the Communist takeover of China. The two got along well and discussed a merger in general terms, but Levin didn't see how it could be structured.[87]

By the time the two spoke again in December 1999, the dot.com boom was in its white-hot, final phase. Technology and telecommunications stock prices were going practically straight up. As far as Wall Street was concerned, it was "Internet or die" when it came to the future of American business.

Working with a very small number of advisors, Levin crafted a deal over the next few weeks with Case and his team to merge Time Warner and AOL in the largest such transaction in business history. When the deal was announced on January 10, 2000, the $183 billion price tag and audacity shocked many in the communications industry, as well as within Time Warner itself. Former AOL stockholders would hold about 55 percent of the combined company's shares, and former Time Warner stockholders would hold about 45 percent of the new company on a fully diluted basis.

Levin was convinced that he had closed out the 1990s by putting his company at the forefront of the Internet revolution and making it a leader in the 21st-century business world. All of Time Warner's divisions—including Time Warner Cable—were going to be winners, his advisers confidently predicted. And as with the Time-Warner merger announced in 1989, Levin was betting that the whole was worth more than the sum of its parts. That wager was put to the test sooner than anyone could have imagined.

"We (Time Warner Cable) were ahead of the crowd with high-speed data, we were ahead of the crowd with video on demand, we were ahead of the crowd with advanced digital, we were ahead with voice, and we were ahead of the crowd with DVRs. We were ahead of the crowd."

— Mike LaJoie, *Chief Technology Officer, Time Warner Cable*

As the world entered a new century, the momentous events to come—inside and outside the company—would frequently test and tax Time Warner Cable employees. Time and again, however, they would rise to the challenge over a decade that would end with the company embarking on an independent path.

Chapter 6

Navigating Change

2000-2011

AOL Time Warner CEO Jerry Levin, left, and co-COO Robert Pittman, far right, celebrate the listing of the combined company's stock in January 2001.

The group of Time Warner Cable executives surveying the January 2000 Consumer Electronics Show exhibits in the Las Vegas Convention Center looked forward to the year ahead. Led by CEO Joe Collins and president Glenn Britt, as well as technology gurus Mike LaJoie, Mike Hayashi, Jim Chiddix, and others, the team had a lot to be confident about. They were among the cable industry's technology leaders, and at the top of their game.[1]

No more scrounging for dollars, as Britt had been forced to do in the mid-1990s to launch Road Runner. The cable team earmarked $2 billion in capital spending for the year, up from an already hefty $1.6 billion in each of 1998 and 1999, "to facilitate a more aggressive rollout of Time Warner Cable's popular digital cable and Road Runner-branded high-speed online service."[2] The company offered digital cable in 30 of its divisions and was on track to roll out the service to the remaining four by yearend.[3]

Network-based video on demand services had been tested in Hawaii; Austin, Texas; and Tampa Bay, Florida, in 1999. In December 1999, video on demand was formally launched in portions of Hawaii. The new service gave customers the ability to pause, rewind, or fast forward[4] *It's a Wonderful Life* or other holiday movies to their hearts' content. Additional divisions followed shortly with video on demand launches of their own.

As the team wound its way through the CES exhibits, Hayashi took a call on his cell phone. "Joe, this call is for you," he said, handing the phone to Collins. Collins paused. "Oh really?" he said into the phone. He ended the call and returned the cell phone to Hayashi.[5] Collins immediately returned to New York.

"Cable's Future Was Wrapped up in the Internet"

With that call, Collins learned that Jerry Levin and his small group of advisors were just days away from announcing the record-setting merger of Time Warner with America Online, Inc. Many of the assumptions he and his team had made about the cable business suddenly seemed as much of a sure thing as drawing a straight flush in one of the nearby casinos. Like the other Time Warner division leaders, Collins had not been privy to the AOL talks until very late in the process. While Levin kept a tight lid on the deal, he made it clear that broadband cable access to the Internet was central to the vision of merging the two communications industry giants.

Steve Case and Jerry Levin, the chairmen and CEOs of AOL and Time Warner, respectively, shake hands on their January 10, 2000, agreement to combine AOL and Time Warner. Regulatory and other concerns delayed the closing of the deal for almost exactly one year.

"My sense was that the Internet and cable were going to be allied," Jerry Levin recalled. "Cable's future was wrapped up in the Internet."[6] Steve Case, AOL chairman, who would become chairman of the combined company, noted that Time Warner "provided an unparalleled array of assets for AOL to transition to broadband."[7] Levin asserted that the deal in turn would spark the "digital transformation of Time Warner."[8]

Searching for Synergy

It took roughly a year for the AOL-Time Warner transaction to win regulatory approval. During that time, different working groups met repeatedly to focus on ways to leverage the strengths of each company. There was reason for optimism, at least at first.

In some ways, it was a repeat of the months following the merger of Time Inc. and Warner a decade earlier. "The AOL deal brought a renewed interest in running all of the company's different pieces on a synergized basis," said Rob Marcus, who at the time was senior vice president for acquisitions at Time Warner. "For example, there were specific mandates to buy advertising on one another's properties," he added.[9]

Marcus also noted that, "Steve [Case] was a huge believer in cross pollination of the assets. I think it's fair to say that his conception of that, though, had AOL in the center, and had all of the other businesses as spokes around that hub."[10]

Time Warner Cable continued to maintain its innovative edge despite the turbulence caused by the AOL deal. The Time Warner Cable team also updated its customer focus to battle rivals in an increasingly competitive marketplace.

"It Wasn't Clear That There Was Any *There* There"

Time Warner Cable chief technology officer Jim Chiddix was a key participant in many of the early AOL-Time Warner meetings where broadband led the agenda. He and other Time Warner Cable technologists were taken aback by what they learned, or didn't learn, sitting down with their AOL counterparts. His reflections were echoed in a slightly different fashion by numerous Time Warner executives:

From a technology point of view, it wasn't clear what AOL was going to be, other than another product to be delivered over cable modems, which it already was. I mean, people were accessing their AOL accounts through cable modems and getting much faster connectivity. And AOL had its own sites, but the Internet had such a rich array of sites that it wasn't clear, to me at least, quite where that led.

What was clear was the clash of cultures. There had been a bit of a clash of cultures 10 years earlier, in the Time and Warner merger, but nothing like this. The first big meetings with AOL were discouraging, in terms of a sense from them that they somehow knew secrets that we didn't know, that they were going to be able to change something in some wonderful way.

It was an arrogance that was difficult, I think. It was difficult for me and for other Time Warner people. Not—hopefully—not just defensively. But because it wasn't clear that there was any there *there. And we sort of never got over that. The* there *never did seem to emerge.*

Whether it could have or not, I just don't know. I still can't tell you what might have made the merger synergistic. It was just one of those unfortunate business adventures. It wasn't much fun, either, because the synergy wasn't there. There weren't fun things to work on together that led somewhere. It was much more of a clash.[11]

Boom to Bust

Within a few months of the announcement, the already tricky business of getting disparate corporate cultures to work together got much more complicated. The dot.com boom went bust beginning in the spring of 2000. Even before the tech-stock-fueled NASDAQ index hit an all-time high in early March, there were clear signs that the mania for all things Internet was cooling, as many Internet "pure plays" saw their stock prices crater. The share prices of AOL and Time Warner also dipped from their highs.

It took McKinsey & Co. consultant Peter Stern three hours driving in a blinding snowstorm from Stamford, Connecticut, to reach Midtown Manhattan on January 18, 2000, for a meeting with Time Warner and AOL executives. The drive was well worth it for Stern, who had worked as a consultant to Time Warner in 1998 and had his

Federal Trade Commission chairman Robert Pitofsky addressed reporters in December 2000 as the FTC, after lengthy consideration, approved the AOL-Time Warner transaction.

first exposure to the cable business while developing a cable business strategy for TWE with Glenn Britt. At the January 18 meeting, he was named the McKinsey engagement manager leading all the teams working on the integration.[12]

Stern approached the assignment convinced that the deal could work. The merger was "one of the great, great opportunities to transform the media and Internet space," he believed.[13] He was assigned to the four-person leadership team of Dick Parsons and Rich Bressler, president and executive vice president of Time Warner, respectively; and Bob Pittman, CEO of AOL, and AOL vice chairman Ken Novak. The foursome's goal was to focus on ways to bring the two companies together. As time went on, however, Stern and many others both inside and outside the companies began to question whether the merged companies would be able to operate together effectively.[14]

Within a few months of the AOL-Time Warner deal being announced, the dot.com boom, already losing steam, went bust. The technology-stock-dominated Nasdaq Stock Index plunged a record 361.58 points on April 14, 2000.

Vice chairman Ted Turner, while rarely acting as a hands-on manager at Time Warner, had been stripped of his management role by Levin, with Pittman set to assume most of his duties. Levin acted within the terms of Turner's contract, but Turner was incensed.[15]

The personal dynamics of the leadership team also raised concerns. Like Chiddix, Stern was taken aback by the inability of key players from both sides to find much, if any, common ground. The fact that the share prices of both companies were headed south didn't calm nerves or egos. By April of that year, it was hard for Stern or others to find much evidence that the integration team was able to function effectively as a group.[16] Stern and his team continued to do the best job possible for their client. His consulting stint over in January 2001 with the merger completed and approved by regulators, Stern returned to his McKinsey office in Stamford. But he was back working with Time Warner, this time on a full-time basis, before he knew it.[17]

As AOL's stock price fell, potential synergies were becoming more elusive. Some may have questioned whether Time Warner should proceed with the deal. Levin wouldn't hear of it.[18]

SPORTS ★ ★ ★ ★ FINAL

DAILY NEWS

50¢ www.nydailynews.com NEW YORK'S HOMETOWN NEWSPAPER Tuesday, May 2, 2000

DONNA HANOVER

SUSAN WATTS DAILY NEWS

Donna scrubs sexy role

Postpones 'Monologues'

SEE PAGE 7

$150M jackpot

Lottery surges in Jersey

SEE PAGE 2

RICHARD DREYFUSS

ANDREW SAVULICH DAILY NEWS

Stars & strike

Actors picket for ad bucks

SEE PAGE 7

Time Warner yanks Disney's WABC/Ch.7 in row over $$$

CABLE MADNESS

SEE PAGES 3-5

In the spring of 2000, Time Warner Cable temporarily ceased carriage of Disney-owned ABC stations in areas including New York City and Houston. The company argued that Disney was demanding too much money for the right to retransmit its programming and refused to agree to Disney's increasingly short-term extension offers. While the stations were back on Time Warner Cable's lineup within a matter of days, more than a decade later retransmission consent regulations continue to weigh heavily on relations between cable distributors and broadcasters.

RIGHT
AOL Time Warner Chief Technology Officer Bill Raduchel holds one of the ubiquitous AOL program discs that were mass-mailed to millions of consumers in the United States.

BOTTOM
The merger of AOL and Time Warner held great promise in combining what was perceived as the best of traditional media and entertainment with the best of the Internet.

AOL Time Warner Merger

Joint Proxy Statement - Prospectus

Notice of Special Stockholder Meetings

Time Warner Stockholder Meeting June 23, 2000 New York, New York	America Online Stockholder Meeting June 23, 2000 Vienna, Virginia

America Online TIME WARNER

ABCs of Retransmission Rules

As if Time Warner Cable didn't have enough to occupy its leadership, a long-simmering regulatory issue suddenly flared up in the spring of 2000. The issue of cable retransmission consent requirements to carry broadcast programming—created by the '92 Cable Act and retained in the '96 Act—had not had much of a lasting impact on the cable industry for the balance of the 1990s. That changed on May 1, 2000. After months of contentious negotiations, and with ABC willing to grant only increasingly short extensions—down to less than a single day at the end—Time Warner Cable saw no other option than to refuse to agree to ABC's conditions and was therefore required to drop the Walt Disney Company-owned ABC stations from Time Warner Cable systems serving more than 3 million homes, including in New York City and Houston. Time Warner Cable argued that Disney was asking far too much consideration for its right to retransmit Disney/ABC signals to cable subscribers.[19]

Time Warner Cable's logic was valid, but its timing couldn't have been worse. The high-profile blackout—actually a "blueout," as that was the color customers saw on their television screens instead of the ABC programming—played right into the hands of rivals and regulators who questioned the pending AOL-Time Warner merger. Time Warner Cable put the Disney/ABC stations back on its systems in a matter of days. But the clash added to the regulatory and public scrutiny of the merger, and the contentiousness between broadcasters and cable operators on the retransmission consent issue continues to the present day.

At the NCTA convention in New Orleans the following week, FCC chairman William Kennard warned that the dispute "raises demons of distrust among consumers" and "concerns about whether or not the cable industry is capable of being the honest gatekeeper to the Internet that it says it wants to be."[20]

From Synergy to Best of Breed

Talk of finding synergy between AOL and Time Warner businesses, especially the AOL dial-up service and Road Runner, resurfaced at times through the year following the January 2001 deal closing and throughout 2002, but little concrete progress was made on the synergy front. Peter

Stern returned to Time Warner as the full-time vice president of strategic initiatives as of June 2001. He followed his mentor, Dolph DiBiasio, who was named executive vice president of strategy and investments. Going forward, the strategy "was to run the businesses as best of breed, so let them succeed in their own right, to integrate in certain areas where it made sense, whether it was for purchasing efficiencies or there was a specific set of initiatives that we would pursue," Stern said.[21]

Even though the public perception both before and after the merger may have been that AOL was the hip new-media company with great national brand recognition and Time Warner was a bunch of suits running old media, the brand equity or value was apportioned very differently. After the merger, AOL's name recognition was acting as an anchor on the company because the name was held in low regard—its mass-mailed program discs were being used as drink coasters. Road Runner "had very high customer satisfaction and high awareness in the Time Warner Cable footprint," Stern pointed out. "It was recognized as the next thing. AOL was recognized as the old thing."[22]

Mystro

The dysfunctional corporate marriage was taking its toll by 2001. Neither Jim Chiddix nor Joe Collins made secret their displeasure with the management and business strategy, or seeming lack thereof, of members of management who came from AOL as it was being applied to the combined company. Chiddix, who had played a leading role in cable industry innovation for more than 15 years, approached Collins in 2001 and said he planned to resign from Time Warner Cable.

Collins instead proposed that the chief technology officer stay with the company, but move over to be the leader of a new technology effort. Collins provided executive leadership. These two, and others, had been discussing what was next on the company's technology horizon, specifically interactive video building on the video on demand infrastructure that Time Warner Cable had been developing.

The new initiative, named Mystro, played an important role in developing many of the video features that Time Warner Cable rolled out as the decade progressed, including "start over" and "look back" functions. Chiddix's team built a prototype of such a system in Green Bay, Wisconsin, within a few years. Acquiring rights to the programming proved a sticking point, as did Time Warner's unwillingness to adequately fund such a stand-alone venture going forward. The group disbanded in 2003, and Chiddix made good on his plans to launch the post-Time Warner Cable phase of his career.[23]

Glenn Britt Named CEO of Time Warner Cable

In August 2001, Glenn Britt, then the president of the company's cable division, received a call from Bob Pittman, co-COO of AOL Time Warner, asking to have a talk over dinner. Pittman's Time Warner counterpart was Dick Parsons. Small talk taken care of, Pittman asked Britt if he wanted to be the CEO of Time Warner Cable.

In August 2001, Time Warner Cable president Glenn Britt succeeded Joe Collins as CEO of the cable company.

After working with Time Warner as an associate principal at McKinsey & Company, Peter Stern joined the company full-time in 2001 and moved over to Time Warner Cable in 2004. He is currently executive vice president and chief strategy officer.

AT&T chairman and CEO Michael Armstrong remained frustrated by the phone company's inability to fully realize the value of its purchase of the cable businesses of TCI and MediaOne in 1998 and 1999, respectively. An unsolicited offer from Comcast for AT&T's cable business in the summer of 2001 led Time Warner Cable to make a bid as well. The AOL Time Warner board would not approve a deal acceptable to AT&T, and Comcast won the bidding.

Britt was fully aware that Collins was unhappy with Pittman's management style in particular, so the offer didn't come as a complete surprise. Britt's performance as president of the cable unit, and his earlier work as CFO of Time Inc. and at HBO, gave him experience that few others possessed. Collins moved over to run the Mystro group. Tom Rutledge, senior executive vice president, became president and chief operating officer of Time Warner Cable.

Parsons also recognized Britt's leadership skills and deep understanding of the cable business and Time Warner corporate culture. "Glenn was the guy [Collins] relied on as his intellectual counterpoint," Parsons said. Britt was "a guy who would go out and structure these deals and who could get his mind around this new emerging area of ... high-speed data and telephony," Parsons added.[24]

Parsons had become familiar with Britt's skills long before. With a background in law and finance, Parsons needed a guide to lead him through the intricacies of the business when he was named Time Warner president in 1995. (They both also served on the ATC board from 1989 to 1991.) Britt was the obvious choice. "Glenn would come over to my office in the evening sometime and just give me a one-hour tutorial on this business, or that business, or how these pieces fit together. ... Glenn was my sensei, my... mentor in the early years in terms of helping me figure out and understand which end was up in a collection of Time Warner businesses."[25]

AT&T Broadband Sale

Even as the management transition was taking place, an opportunity to fundamentally reshape the Time Warner Cable business and place it firmly at the top of the industry ranks landed at Time Warner's door. AT&T proved unable to shake its utility mindset under CEO Michael Armstrong. It acquired TCI in 1998 and the MediaOne cable systems in 1999, but appeared to have underestimated the operational issues plaguing many of the systems it purchased for a total cost of nearly $80 billion, after the sale of certain assets. It was hemorrhaging cash in 2000, and its stock price seemed to sink further by the week, partly as a result of the dot.com bust. Armstrong finally created a tracking stock for the cable assets as well as three other divisions in late 2000, with the intention of spinning off the cable systems as a standalone company.

Comcast had lost out to AT&T in the bidding for MediaOne. It was still on the prowl for a major acquisition, and in July 2001, after private talks with AT&T went nowhere, Comcast made an unsolicited bid for AT&T's cable business. Comcast saw an opportunity to bid on many of the assets it had lost out on to AT&T at sharply reduced prices. It wasn't alone.

AOL Time Warner and Cox, encouraged by AT&T, jumped into the bidding fray that fall. Levin, Collins, and Britt were of one mind when it came to the AT&T cable systems: Big cable deals like this were few and far between. They were eager to bid. After all, the Newhouse and CVI systems that Levin had been widely criticized for adding in 1995 and 1996 now accounted for a significant amount of Time Warner Cable's cash flow.[26] Britt was even willing to temporarily step aside as CEO for a year or two in favor of bringing back Collins, who was better known to the financial community, in order to win the AT&T systems.[27]

It was a tempting deal. Adding AT&T's cable subscribers would give Time Warner Cable control over nearly one-third of the 73 million U.S. cable television subscribers at the time.[28]

"A Missed Opportunity"

Although Britt and others heard from sources inside AT&T that AOL Time Warner had submitted the highest sealed bid for the AT&T cable systems by late 2001,[29] it was Comcast that came away with the AT&T systems, paying $47 billion in cash and assuming $25 billion in liabilities.[30] The Time Warner Cable bid was hampered by a complex plan for ownership of the assets, the company's insistence on a liability cap, and an AOL Time Warner board that was not keen to repeat the grueling regulatory approval process it had just gone through in connection with its own merger. Ultimately, the AOL Time Warner board would not approve a deal that was acceptable to AT&T.

It was a tough loss for the Time Warner Cable team to take. From their point of view, it was yet another sign that the AOL side of the company simply didn't understand the dynamics of the cable business. In a ranking that remains in place to this day, Comcast became the nation's largest cable operator, and Time Warner Cable number two. "I still think of it as a huge missed opportunity," Britt said. "The whole trajectory of the industry might well have been different if we had followed through with that purchase."[31]

Changing of the Guard

The unwillingness of the AOL Time Warner board to zealously pursue the AT&T cable systems helped Levin realize that he was at a turning point in his professional career. He had been under mounting pressure all year from Wall Street, as well as AOL Time Warner chairman Steve Case and vice chairman Ted Turner, to set more realistic performance targets and in general do a better job managing investor expectations about AOL Time Warner's share price. As the largest shareholder, Turner had made no secret of his ire at seeing his net worth drop by several million dollars a day for much of 2001.[32] Turner had also never forgiven Levin for removing his operating responsibilities in the wake of the merger with AOL.[33]

Levin also had personal concerns. His son Jonathan, a New York City public school teacher, had been brutally murdered in 1997.[34] And the horrific events of September 11 left him shocked and reassessing what he and the company ought to be contributing to society. He was incensed, for instance, that Case insisted on holding a previously scheduled AOL Time Warner board meeting shortly after 9/11, and remained so even after Case agreed to hold the meeting over the phone.[35]

AOL Time Warner co-chief operating officer Dick Parsons succeeded Jerry Levin as CEO of the company in 2002, and in 2003 Parsons added the chairman title following the resignation of Steve Case, leaving the company clearly in the hands of Time Warner veterans.

Bob Pittman, the former AOL CEO who served as co-COO of AOL Time Warner with Richard Parsons, resigned in 2002 in the wake of an accounting scandal on the AOL side of the company, though he was not personally implicated.

Apparently no longer able to count on the unqualified support of at least half of his board, nor Case and Turner, Levin announced his impending retirement on December 5, 2001. He invoked a clause that he had inserted into his last contract, following his son's death, which enabled him to retire with six months' notice. At 62, it was time to get on with his life.

The move came as a surprise to many inside the company as well as outsiders. As one of the principal architects of the biggest corporate merger of all time, Levin had been expected to stay in his role at least until his contract expired in 2003. But in the act of leaving, he took steps to ensure that a Time Warner veteran remained at the corporate helm.

As part of his plan to retire early, Levin won board support for naming co-chief operating officer Dick Parsons CEO-designate of AOL Time Warner. Parsons was a popular consensus choice. Whether fair or not, his in-house rival, Bob Pittman, was tagged with bearing responsibility for the dramatic decline in AOL's business. Pittman reported to Parsons as the sole chief operating officer.[36]

Bottoming Out

By the time Parsons officially assumed the responsibilities of CEO at the May 2002 annual meeting, AOL Time Warner's fortunes had gone from bad to worse and were headed toward horrible. In May of that year, the company had disclosed that it was taking a $54 billion charge—the largest such charge ever—to write down what in accounting terms is euphemistically called "goodwill," or the extent to which a deal's price exceeds the value of the assets on a company's books.[37] Parsons also adopted a much more realistic approach with Wall Street compared to Levin and stated that he wasn't going to make any promises he couldn't keep.

Parsons made it clear internally that the overarching focus on synergy was a thing of the past. The "best of breed" strategy received Parsons' official imprimatur.[38] And with that admission, Time Warner division heads felt free to disregard much of the direction they were getting from Case, Pittman, or other executives from the AOL side of the house.

In a meeting in the spring of 2002, anti-AOL sentiment boiled over, as described in former *Washington Post* reporter Alec Klein's book on the collapse of the merger, *Stealing Time*. Jeff Bewkes, chairman and CEO of HBO, cut off a rosy scenario being presented by Case. "Every one of us is growing, making the numbers. The only problem in this construct is AOL."[39]

A few months later, the company's stock price hit a low of under $9 per share following news of unorthodox business practices at AOL that, according to Klein and others, had the effect of inflating revenues. Many Time Warner insiders were outraged, alleging that the practices had inflated the price tag for the original merger. This was just the beginning of the problems leading up to allegations of accounting improprieties at AOL, an SEC investigation, stockholder lawsuits, and financial restatements.

Turner was still the largest shareholder in AOL Time Warner. He had watched aghast as his net worth, once roughly $10 billion, had plunged to about $2 billion by the summer of 2002. He had lost $8 billion in 30 months.

By January 2003, Steve Case, having little support among the executives running AOL Time Warner's divisions, announced his resignation as chairman. He was replaced by Parsons. By month's end, Ted Turner also resigned as vice chairman.

Case resigned almost exactly two years after the merger had been approved. That same day, AOL Time Warner announced it would write down another $45 billion in goodwill on its books. That brought the value destroyed by the largest merger in history to just shy of $100 billion.[40] The company dropped AOL from its name in October 2003.

"Doing the Best for the Customers"

Veda James
Retail Sales
Austin, Texas

My first job at the company starting in 2000 was as a customer service rep, but in September 2008 I became a retail sales rep. I get a chance to talk to customers more and understand what they want. It gives me a chance to be flexible and take more control during interactions with them. I have more time to tell customers about our products and services, and the things they'd like to have in their homes. It gives me a greater sense of ownership because of this direct contact.

We all work together very closely. Ours is a small team of six. Two team members had babies, one week apart, in 2009, and then three others had babies in 2010, all around the same time. We were short staffed, but we worked together and pulled it off... skipping lunch, working long hours, not taking vacation. It was a test of patience and endurance, but we covered for each other. We genuinely care for our team members, and this is something new to me. I never did this before or faced a challenge like this in my previous jobs! Integrity in everything we do is a very important value. It means doing the best for the customers, being honest, and treating them with respect.

There was something I did in 2009 that makes me very proud. A customer had by mistake left behind a lot of cash, a couple of checks, and other papers at our center. I found the money and, with my supervisor's help, tracked down the customer. He was so thankful. He kept saying over and over, "I didn't think I'd ever see that money again."

Advance/Newhouse to Bright House

By 2002, Bob Miron of Advance/Newhouse wanted to make sure that its arrangements with Time Warner Cable were in his company's best interests. "We weren't sure whether AOL Time Warner was more interested in developing the cable systems and making them grow or seeing AOL grow instead of the cable systems," Miron recalled.[41]

When Advance/Newhouse formed the TWE-A/N partnership with Time Warner Cable in 1995 covering nearly 4.5 million subscribers, Miron, as its key negotiator, had very deliberately inserted a "divorce" clause in the partnership agreement. It said that regardless of who initiated the break-up, Time Warner Cable would divide the systems in the partnership into three groups—one based on systems in the Carolinas, another on the upstate New York systems, and the third comprising the Florida systems. Advance/Newhouse would get to pick one of the groups, reflecting its one-third ownership.

It was typical of Miron to structure an agreement that relied on interlocking trust: "In doing the transaction and convincing Joe and the others that this was fair, I always said to them, 'Listen, if you divide the piles evenly, then you won't care which pile I pick first. If you try to fool me and I catch you, I'm going to win. If you try to fool me and I don't catch you, you win.'"[42]

As Advance/Newhouse's concerns about AOL Time Warner's direction intensified, Miron began to consider triggering the TWE-A/N "divorce" clause. The parties came up with another approach, however; one that would allow Advance/Newhouse to stay in the Time Warner Cable family but give the valued TWE-A/N partner the right to manage the day-to-day operations of one of the three system groups mentioned in the divorce clause.

Miron chose to take on day-to-day management of TWE-A/N's Florida systems, most of them centered in the Tampa and Orlando markets, and renamed the operation Bright House Networks. The systems were held (and remain today) in a subsidiary of TWE-A/N. "The partnership exists, and we work together in a number of areas, including programming, engineering, and regulatory," Miron said. "And we still meet on a monthly basis."[43]

Restructuring TWE

Even as the company was enduring wrenching change at the corporate level, AOL Time Warner took the opportunity to simplify the TWE partnership structure containing most of its cable assets. As a result of its purchase of AT&T's cable assets, Comcast had acquired MediaOne's 25.5 percent stake in TWE. That left Comcast with cable assets valued at $10 billion or more tied up in a partnership it didn't control and, as a result of the consent decree it entered into to get regulatory approval of its AT&T deal, Comcast was required to divest it ownership interest in Time Warner Cable no later than 2007.[44] AOL Time Warner and Comcast agreed that the TWE arrangement should be revised.

In connection with the restructuring of TWE in March 2003, AOL Time Warner acquired full ownership of TWE's content businesses, including Home Box Office and Warner Bros., as well as TWE's interests in the WB Television Network, Comedy Central, and Court TV. All of AOL Time Warner's cable assets, including those in TWE, were transferred to the newly created subsidiary Time Warner Cable Inc., in which AOL Time Warner held a 79 percent economic interest and of which TWE became a subsidiary.[45]

In exchange for its previous interest in TWE, Comcast, in turn, ended up with a 21 percent economic interest in Time Warner Cable Inc. It also received AOL Time Warner preferred stock convertible into $1.5 billion of AOL Time Warner common stock, and $2.1 billion from Time Warner Cable as repayment of a preexisting note.[46]

"Smarter Marketing"

As president, Britt had been taking stock of the Time Warner Cable business and its competitive position, and he continued his assessment in earnest on being named CEO in 2001. He didn't like what he saw. "It became brutally apparent to us that we had to get a lot better at competing" with satellite and telecommunications company rivals, he said. Time Warner Cable wasn't alone among cable operators in this regard, of course, but that was cold comfort as far as Britt was concerned.

The company needed to react and move faster in response to changing market conditions. It needed to be more innovative and launch new products more often than every four or five years. "And most important, I thought that we really needed to be much more sophisticated in our marketing," Britt said.[47]

The company beefed up its management ranks in support of a more aggressive approach to marketing. John Billock, president of HBO's U.S. Networks, joined the cable team as vice chairman and chief operating officer in October 2001. Tom Baxter had joined Warner Cable in 1980 and rose through the ranks, leaving in 1990 to become president of Comcast. After three years with an investment firm and a dot.com company, he returned to the fold, joining Time Warner Cable as president at the same time as Billock. Tom Rutledge resigned and joined Cablevision, where he became chief operating officer.[48]

Time Warner Cable had enjoyed modest successes with direct marketing and door-to-door sales, but the efforts were hit or miss. "We didn't know much about our consumers. We didn't have big databases like a lot of direct marketing companies have," Britt said.[49]

"We were decentralized, so not only our marketing but our advertising was very uneven. It was pretty good in some places and really bad in some other places, and it was done city by city," Britt added. "So I set out to have what I called 'smarter marketing.' Some of our marketing people took offense to that because they thought I was saying that they were dumb, but I really meant we had to be more sophisticated."[50]

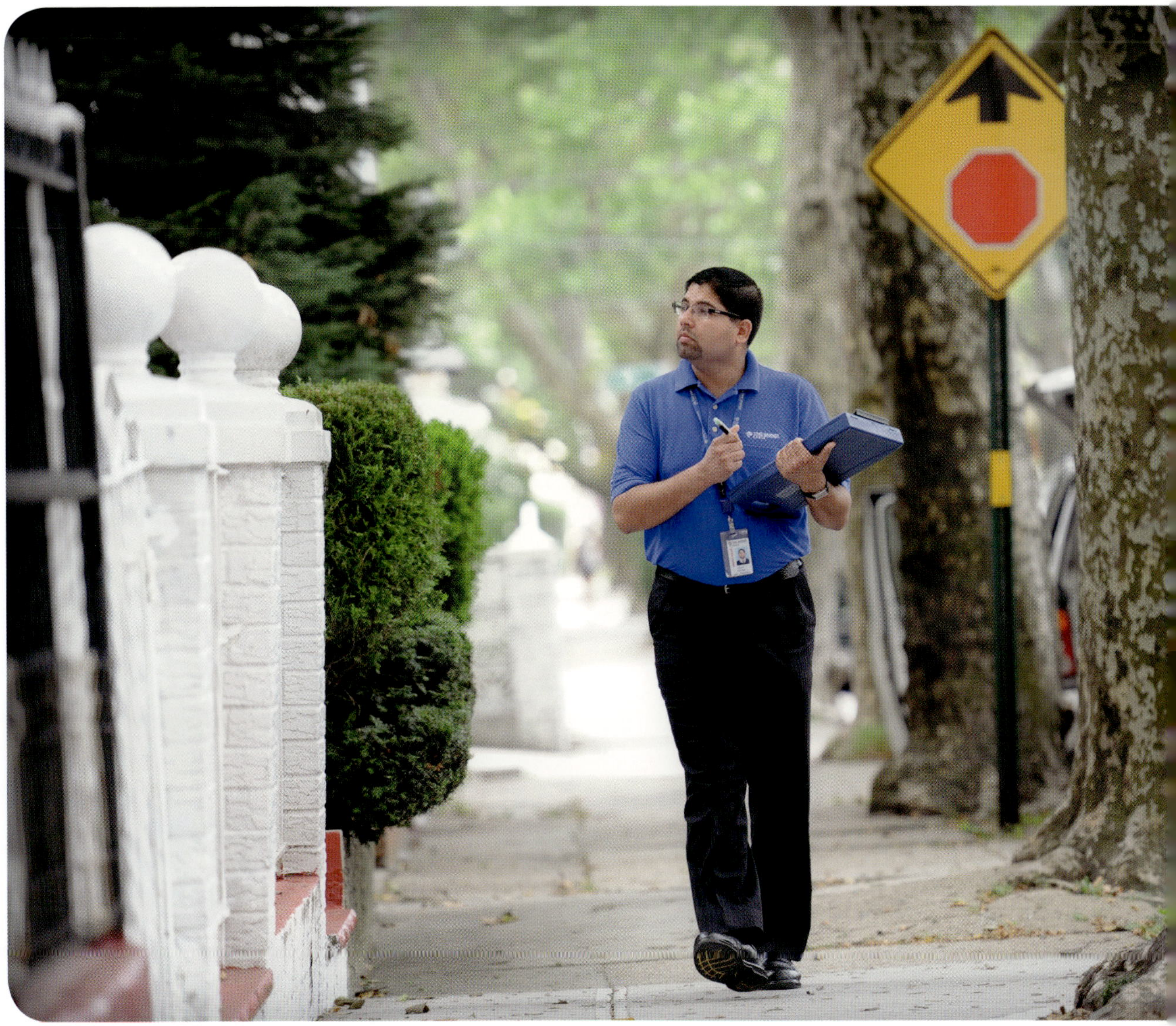

Sending out direct-sales staff in markets such as Jamaica, Queens, was just one of the ways in which newly named CEO Glenn Britt sought to drive a much more competitive culture throughout Time Warner Cable shortly after taking the company reins in 2001.

9/11: "We Realized for the First Time How Critical Our Service Really Is."

Time Warner Cable's NY1 acts as the eyes and ears of New York City, and never more so than during the critical days and weeks following the 9/11 terrorist attacks on the World Trade Center.

Time Warner Cable employees were shocked by the tragedy that struck on the morning of September 11, 2001. Nanette Posman, communications manager in New York City, can't forget the "relentless sounds of sirens" outside the 23rd Street offices in Manhattan "and the strangely quiet fear" within.[51] Marion Boykin, on the company's government relations team in New York City, stood dumbstruck at the corner of 23rd Street, staring at the flaming towers framed by the view facing south down Fifth Avenue.[52]

In Flushing, Queens, one customer service representative nearly fainted as she realized that a close friend had been at work in one of the towers and perished in the collapse. On the roof of the Flushing building, workers gathered to stare at the smoke streaming from the World Trade Center towers. A photographer working for the company caught the horrific sight on film.[53]

Shock quickly gave way to determined action. Employees at all levels channeled their anxieties and fears into focusing on what Time Warner Cable could do to help the city respond to the unprecedented crisis. With traffic in and out of Manhattan banned, many had to walk to work from the outer boroughs. One of the most immediate services the company provided was to keep its products up and running (except in the immediate vicinity of the attacks) while over-the-air television signals relayed from the giant antenna atop the World Trade Center were knocked out.[54]

Less than 24 hours after the towers collapsed, desperate officials from Police Plaza in lower Manhattan were on the phone. Their DSL lines were down. They needed Road Runner service, and they needed it ASAP. Barry Rosenblum, Time Warner Cable New York president, made it clear that this and similar requests to follow were top priorities: "Just get it done." Even though the immediate system infrastructure near police headquarters had yet to be upgraded, company technicians worked feverishly to get a router installed that enabled 250 computers to tap Road Runner's broadband services.[55]

In the days that followed, similar requests from City Hall, the Armory, Red Cross facilities, police command trailers, the coroner's office, and the Family Assistance Center on the Hudson River waterfront were fulfilled. The police stopped traffic so technicians could pull cable across the West Side Highway to the center, stringing it from makeshift poles. Later, the company installed 4,000 feet of cable on Staten Island at the landfill that served as a repository for the wreckage from Ground Zero.[56]

For customers in lower Manhattan forced to vacate their homes, Time Warner Cable gave them immediate credits on their bills. As an indicator of how a sense of giving permeated so much of the response to the tragedy, several customers said they would donate their refund checks to the Red Cross or other charities helping victims of the attacks.[57] Time Warner Cable employees in New York and across the country made donations as well that, matched by corporate funds, totaled $1 million by yearend 2001.[58]

The events and aftermath of September 11 changed everything. That included the way New York City viewed Time Warner Cable, and the way company employees thought of their own company. "I think the perception really changed about our role and the significance we played as a company," said Karen La Cava, manager of public affairs. "A lot of us went from thinking we provide entertainment to realizing for the first time how critical our service really is."[59]

Added Harriet Novet, regional vice president, public affairs, "This disaster really brought out the best in our employees. And in our company."[60]

Time Warner Cable employees in New York City, including those staffing this service center in Flushing, Queens, worked above and beyond their job descriptions and workday hours to meet the needs created by the unprecedented crisis in the wake of the 9/11 attacks.

The company-wide rollout of Voice over Internet Protocol (VoIP) phone service that began in 2003 proved a huge hit with consumers and added a significant new revenue stream for the company.

VoIP

Gerry Campbell was sitting in on a Time Warner Cable meeting in Stamford shortly before Christmas 2001. The former Warner Cable executive had returned a few years earlier from Britain, where he had played a leading role in implementing TDM (a switch-based phone system) service for Comcast in the UK. A subsequent Comcast attempt to bring VoIP to the States went nowhere, leaving Campbell open to new challenges. When Comcast's Tom Baxter rejoined Time Warner Cable in 2001, he recruited Campbell to follow suit.[61]

At the Stamford meeting, a Time Warner Cable executive was diagramming on a whiteboard how the company was going to get into the VoIP phone business. To put it nicely, Campbell said, "He didn't know what he was talking about." Campbell sat through the presentation, then went up and politely corrected his whiteboard.[62]

"Can you hold on a minute?" the presenter asked Campbell excitedly. He disappeared, came back a few minutes later, and said, "We want to bring you upstairs, and you tell the guys upstairs what you just told me." Campbell gave the same explanation to the executives upstairs. Christopher Bogart, then president of the company's Time Warner Cable Ventures new-business unit, offered Campbell the opportunity to run the VoIP phone business on the spot. Baxter wanted Campbell to stick with him, but the phone offer was too good to pass up. Campbell was named senior vice president, voice, and charged with rolling out phone service in Time Warner Cable's service areas across the United States.[63]

"All You Can Eat"

Campbell spent 2002 assembling a team and working out the technical details of carrying phone calls over the hybrid fiber coax cable system. A key consideration was how calls over the cable systems would connect with the existing phone network. Time Warner Cable decided to partner with outside vendors to manage that and other crucial parts of the system. "We needed somebody to connect us to the world. We couldn't connect at that time; we just weren't advanced enough," Campbell said.[64]

He focused on the existing limited VoIP test market in Portland, Maine, as the obvious starting point for a nationwide rollout of the new product. Carl Rossetti, who replaced Bogart as the executive in charge of the company's new ventures, and who had run the cable system in Portland for several years earlier in his career, also wanted to launch the rollout there. It was a small enough market so that if they ran into problems, it probably wouldn't make much of a splash in the national media. "If something blows up in Maine," Campbell said, "people are just going to say, 'Did it impact the lobsters?'"[65]

Campbell chose Pine Tree Networks, a small independent phone company in Portland with only 3,000 customers, to provide the cable company with a link to the nationwide phone network. Even though Campbell was a native New Englander, the locals were suspicious of his motives and the multi-billion-dollar corporation he represented. "I used to sit across the table, and they'd say, 'Why are you doing this? Why don't you just buy us?'" Campbell recalled.[66] Campbell and his team convinced them that the relationship would be mutually beneficial. In another aspect of launching phone, Campbell explained that due to the uncertain regulatory outlook, Time Warner Cable wasn't sure that it should be considered a telephone company. Even though it kept state telephone regulators abreast of its plans and provided emergency 911 service, from a regulatory point of view it wasn't clear whether it was providing a regulated telephone service or a high-speed data service.

Consistent with the way Time Warner Cable priced video offerings, Campbell and Rossetti wanted to offer a single, "all you can eat" phone package that included local and long distance calls. Their research suggested that customers wanted this kind of simplicity and that the company could price the product to meet anticipated long-distance usage and still turn a profit. They settled on $39.95 a month. It was a gutsy call, and one that soon had national phone companies scratching their heads in wonder at the cable executives.

They chose January 2003 for the test of the new VoIP package in Portland. Campbell and Rossetti budgeted for 1,500 phone customers in the first year. More than twice that many signed up in the first *month*, equal to the total number of traditional phone customers using Pine Tree's phone service. From the field, it clearly looked like Time Warner Cable had a hit on its hands.[67]

"Like Changing the Fan Belt with the Engine Running"

In order to roll out the service to all Time Warner Cable divisions beginning in the second half of 2003, Campbell and Rossetti engaged in lengthy discussions with AT&T, Sprint, and MCI to provide on a nationwide scale the connection and other services they were getting in Portland from Pine Tree. AT&T simply wasn't willing to shoulder the risk that the cable company had accurately pegged the average customer's long-distance calling habits in its $39.95 all-you-can-eat package. MCI, on the other hand, had been offering a flat-rate plan, and so was more comfortable with the one-rate practice. So as not to be tied to a sole provider nationwide, Time Warner Cable added Sprint Corp., and then split its divisions across the country, with roughly half using MCI and the other half using Sprint.[68]

Campbell and his team executed the rollout strategy as rapidly as possible. Virtually all the divisions were offering phone service by the end of 2004.[69]

Taking the strategy used for the national rollout of Road Runner to the next level, and so as not to disrupt existing cable service, they effectively franchised the phone service to all their divisions. "The concern was that this is like changing the fan belt with the engine running. We didn't want to impact the core business. We didn't want to take the core business down," said Campbell.[70]

"We had 31 divisions ... and so we hired 31 vice presidents of digital phone, and we brought in basically all the people from the phone business, because we needed to insert the culture into the cable culture and convert it in order to understand phone."[71] Project managers stayed in the divisions for nine to 10 months after the launch, five days a week, to ensure that the rollout went as smoothly as possible.[72]

What Would Verizon Do?

While Campbell was working in Portland, Rossetti recruited Peter Stern in Stamford to focus on Time Warner Cable's phone strategy vis-à-vis what other cable and phone companies were planning. To analyze how Verizon executives would view the opportunity, Stern had Glenn Britt play the role of Verizon CEO Ivan Seidenberg and work through the analysis of how Verizon would approach VoIP phone service. Even if Time Warner Cable didn't roll out its VoIP phone service, Britt and Stern concluded that despite the tremendous expense involved, Verizon would almost surely decide to go ahead and run fiber-optics to offer phone and broadband Internet service.

"Our conclusion was Verizon's going to deploy fiber no matter what we do," Stern said. "Therefore, our strategy should be to deploy phone as fast as possible and gain a couple of years on them so that we have a triple play before they do." By the spring of 2003, the company had decided that it should aggressively roll out phone service, with the target of having it available to customers in all divisions by the end of 2004.[73] As predicted, Verizon aqqressively pursued its fiber to the home strategy and launched its FiOS broadband and VoIP phone service in September 2005, more than two years behind Time Warner Cable.

Time Warner Cable technicians answered the technological challenge presented by the introduction of VoIP phone service and signed on for advanced training courses, some offered by retired AT&T phone technicians.

"We Retooled Everybody"

As phone vice presidents were put in the front offices of Time Warner Cable divisions across the country, technicians needed to be brought up to speed on the new technology as well. After Portland, Maine, Campbell and his team went to Charlotte, North Carolina, to launch VoIP service. To educate the teams of cable frontline technicians on the phone business, Campbell turned to a group called the AT&T Pioneers, comprised of retired phone company workers. "I brought in an army of retired phone guys who went out and crawled under houses in the Carolinas at 50-something to 60-something years old, and showed our guys how to do cable TV and phone offerings together. Taught them how to troubleshoot, what tools to use. And we retooled everybody. We retaught them how to become phone people."[74]

Triple Play

On the marketing front, Sam Howe arrived in Portland in 2003 to help lead the phone rollout. He had previously worked at Telewest in Great Britain, where he was one of the early architects of the cable "triple play." In the late 1990s, Time Warner Cable had positioned its early trials of phone service as a secondary line, concerned that consumers weren't willing at that point to consider replacing their existing phones with one that worked over the cable connections. Yet when the company began widely rolling out VoIP in 2003, the service was marketed as "Digital Phone," positioned in the market as an "important complement to video and high speed data in helping make Time Warner Cable the provider of choice for consumers," said John Billock.[75]

The idea of combining the three services in a package, while part of Time Warner Cable's long-term plan for phone service from the start, didn't fully gain traction until early 2005. That year, Howe was promoted to chief marketing officer, merging what had until then been three distinct marketing groups with 66 employees—one group in cable, one in Road Runner, and the newly created phone group. "What drove us together was not only my being appointed," said Howe, who left Time Warner Cable in 2011 and was succeeded as chief marketing officer by Jeff Hirsch. "It was the idea that we had to create a bundle and make the bundle work. Bundling really came to the fore ..."[76] Chuck Ellis, a long-time Time Warner Cable senior executive who had most recently run the company's cable marketing efforts, retired that year.

The launch of digital phone and the triple play package resembled the sort of moonshot success not seen since the early days of HBO on satellite 30 years earlier. From 2004 through 2006, the number of triple play customers soared more than tenfold, from 145,000 to 1.5 million.[77] In 2007, Time Warner Cable's triple play ranks swelled by more than 50 percent to 2.4 million, and its VoIP subscriber base jumped to nearly 3 million.[78] "We got to market quickly," said Peter Stern. "We had a real time-based advantage, and it reflected itself in faster subscriber growth and profitability growth than some of our peers during a critical period prior to the availability of fiber to the home through the phone companies."[79]

"Our More Competitive World"

The nationwide rollout of digital phone reinforced in the minds of Britt and other senior executives that the company's widely dispersed, decentralized divisional structure no longer matched what was becoming an increasingly uniform series of product offerings. Customers in one region were being offered essentially the same package of products and services as in another—local channels in one market substituted for local channels in another. The divisional structure, which had been created in the 1980s so that the company could keep its operating executives in touch with local trends and needs, no longer seemed as vital. The key was figuring out how to standardize functions that could be combined to deliver significant economies of scale, while still retaining aspects of the local focus that consumers have long expected from the cable companies that serve their communities.

The plethora of divisions made it harder for Time Warner Cable to leverage many of its strengths. "We really weren't using [optimal advantage of] the scale we had to be great at marketing, to be great at engineering, or even great at purchasing things," said Britt. "In our more competitive world, where we were competing against people with bigger footprints, notably the phone companies and satellite, we really needed to be much more cost efficient than we had been," he added.[80]

Operating in a more competitive arena, Time Warner Cable also couldn't be as tolerant of underperformance. A slump in business, in which

Time Warner Cable lost customers in the first nine months of 2004, triggered a management review of operations. That review concluded, among other things, that the role of president and chief operating officer should be combined. As a result, the two executives holding those positions, Tom Baxter and John Billock, retired.[81]

After a nationwide search for a single executive to fill that post, Britt in 2005 tapped Landel Hobbs to serve as chief operating officer. In 2001, Hobbs had joined Time Warner Cable as chief financial officer from Time Warner, where he had served for a year as vice president of financial analysis and operations. Prior to that, he had spent seven years with Turner Broadcasting in a series of financial positions. Hobbs became the lead architect of creating a regionalized operating structure throughout the company.

Landel Hobbs was named Time Warner Cable chief operating officer in 2005 and oversaw the consolidation of the company's operating divisions before leaving in late 2010.

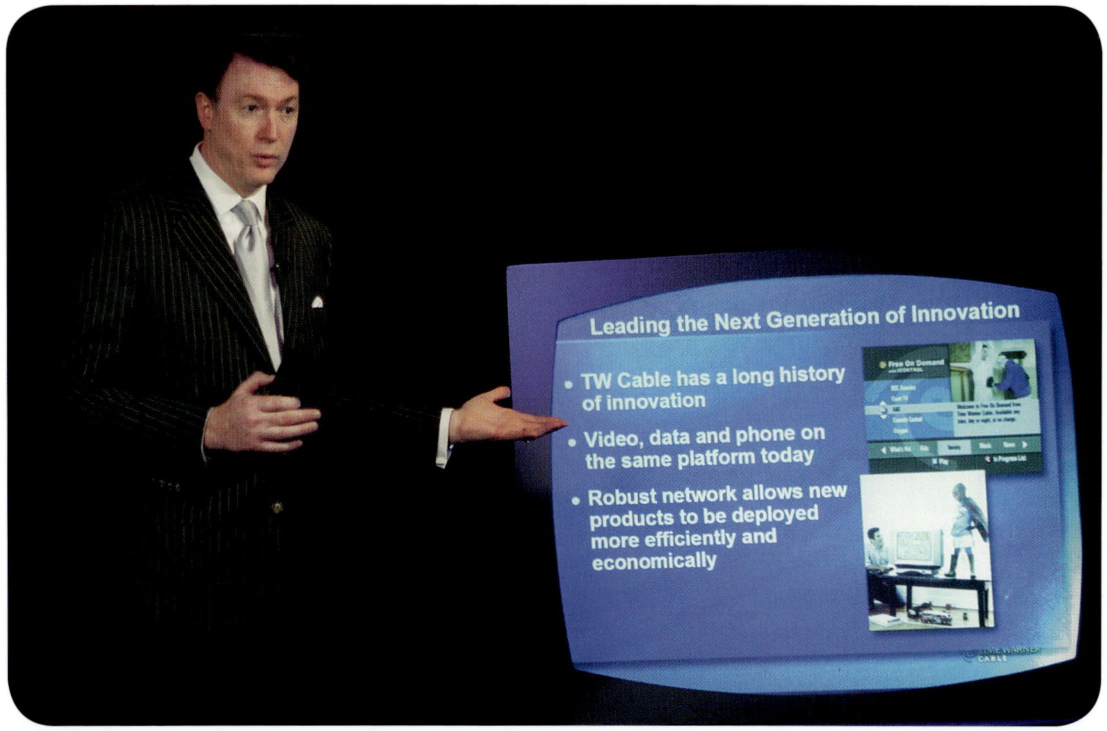

"It Is Change Itself that Continues to be the Most Interesting Part of My Job."

Paula Smith
Regional Director
Employee Communications
Corpus Christi, Texas

Paula Smith joined Time Warner Cable in Corpus Christi in 2001 in a newly created role to help boost community relations. Since 2001, there have been "changing priorities, changing technology and changing organizational structures, and it is change itself that continues to be the most challenging and interesting part of my job." She joined the industry when it "exploded with new technology and new competition." Since then, "the explosion continues, and as a company, we are a major part of it."

Paula has many stories about how Time Warner Cable employees support each other and work together—and none is more riveting than her own. "When my son was deployed to Iraq, employees filled so many care packages I had to borrow a truck to take it all to the post office. When the unthinkable happened and he was killed in action, my peers quietly ensured that I learned to function again."

As her role changed and expanded, Paula's responsibilities increased. Now as regional director, employee communications for Time Warner Cable's West Region, "I work with my counterparts across the region to ensure employees have the information they need to continue to do the best job possible."

Glenn Britt reached into the Time Warner ranks to bolster his executive team in 2005 by naming John Martin, left, head of investor relations at the parent company, chief financial officer of Time Warner Cable, and Rob Marcus, head of Time Warner mergers and acquisitions, as senior executive vice president.

Time Warner Cable brought in two additional senior executives from Time Warner in 2005. Robert Marcus, who had been head of M&A at Time Warner and wanted to broaden his scope of responsibilities, joined Time Warner Cable as senior executive vice president. John Martin, head of investor relations at Time Warner, joined Time Warner Cable as chief financial officer to fill Hobbs' position. Two years later, Martin returned to the parent company to be CFO following the retirement of Wayne Pace, who has served as a director of Time Warner Cable since 2003. Marcus, in turn, took the CFO reins at the cable company.

To outsiders, Marcus appeared to be on an odd career path. "Based on résumés, it didn't make a lot of sense," Marcus conceded. "But based on culture and working relationships, and Glenn's confidence in me, it probably did make sense. And it was a smooth transition."[82]

Hobbs worked closely with Britt to begin the mid-decade transition to an organization focused on six regions rather than more than 30 divisions. The Carolinas region, where digital phone was rolled out on the heels of the Portland launch, served as a laboratory of sorts to see what worked best in the creation of a region. It was important, the company felt, that the regions not simply be an agglomeration of mini-divisions stuck together. Real economies of scale and focus needed to be achieved.[83]

Fred Dressler: The Dean of Cable Programming

In 2007, the cable industry lost one of its leading programming pioneers when Fred Dressler, who after 30 years had retired from Time Warner Cable the previous year as executive vice president of programming, succumbed to pancreatic cancer. Dressler joined ATC in 1976 from a Denver broadcast television station at the urging of company co-founder and industry legend Bill Daniels. He started out managing cable systems for ATC, then moved to company headquarters in Denver, where he made a name for himself in programming.

Dressler was known as a tough negotiator and a consistent critic of rising programming costs, especially in sports programming. He also championed unique content and niche programming during the 20 years he headed TWC's programming operations. He was a founder of E! Entertainment, Sun Sports Network in Florida, SportsNet New York, and the In-Demand Network. In addition, Dressler constantly advocated new ways of delivering video content to consumers and prompted Time Warner Cable to speed its introduction of video on demand, DVRs, and high-definition programming.

"Fred's reputation as a fierce negotiator was always tempered by his humor and sense of fairness," CEO Glenn Britt said in a note to employees following news of Dressler's death. "His legacy will be felt for many years, not just in our programming department, but throughout our organization."[84]

Carol Hevey started at Warner Amex Cable in Columbus, Ohio, as a secretary in 1982 and quickly rose through the ranks, later serving as division president of Time Warner Cable operations in areas including Milwaukee, Boston, and Portland. In 2005, she was executive vice president in charge of the company's Carolinas operations, which served not only as the laboratory for a regionalized structure and one of the earliest digital phone markets but also as test beds for other new services and features the company introduced throughout the mid-to-late 2000s, including Start Over® and DVRs. These services, Hevey recalled, represented a turning point in how customers experienced cable. They put more control than ever in consumers' hands. Along with other video on demand offerings, these services gave viewers their first taste of being able to create their own programming schedules. "I remember so many people—employees and customers—who, when you explained to them what the DVR was, would say, 'Well, I have a VCR, and I don't record anything because I can't figure out how to program it; I would never use that device,'" she said. "And you give it to them, and within days they're going, 'I don't know how I ever lived without this; this has changed my life.'"[85]

Hevey and her teams throughout the Carolinas demonstrated that a regionalized structure could be successful in creating economies of scale and efficiencies. That success suggested that further consolidation to two regions would yield greater savings and an even greater focus on improving customer service. In 2010, Hevey and Bill Goetz, an executive vice president of operations who joined the company from Comcast in 2002, were tapped to run the East and West regions, respectively.

Carol Hevey, executive vice president of operations, East region, oversees system operations in the eastern half of the country and works closely with her counterpart in the West, Bill Goetz, to boost customer satisfaction, business performance, and employee best practices across the organization.

Their staffs include regional management in engineering and network operations, marketing, communications, finance, and other functions, as well as area vice presidents in charge of the company's technical operations and customer care in local markets. "That frees them up to dive much deeper into the day-to-day operations to make sure the customer experience every day is better," Goetz said.[86] The multi-year process of transitioning to two regions has had some "difficult aspects to it," Hevey acknowledged. "This is such a dynamic business that we have to be flexible; we have to be able and willing to change in order to remain competitive. And one thing we've all learned is that we can change."[87]

TOP
The 2004 conviction of Adelphia Communications' founder and CEO John Rigas, left, on multiple conspiracy and fraud charges was one of a series of steps that led to a joint bid in 2005 from Time Warner Cable and Comcast for Adelphia's systems, which netted TWC about 3.5 million new subscribers and unwound Comcast's minority stake in TWC.

Adelphia Communications

Even as Time Warner Cable was engaged in reconfiguring its national footprint, it seized the opportunity to add more systems, further concentrating its clusters of systems and improving its competitive position in the industry. Britt and his team rued the fact that AOL Time Warner let the AT&T cable systems get away. They weren't going to be thwarted again if another major systems operator came on the market.

By 2004, Adelphia Communications, at its height the sixth-largest cable systems operator in the country, was restructuring under bankruptcy protection and was clearly coming on the auction block. The company collapsed into bankruptcy in June 2002 shortly after it became known that the Rigas family, whose patriarch John had founded the company, owed Adelphia $2.3 billion in off-balance-sheet loans. Despite the gross financial mismanagement at its Coudersport, Pennsylvania, headquarters, however, the company had several attractive cable systems around the country, even if they did need a little TLC.[88]

"There were numerous assets within the Adelphia company that were geographically interesting to us because of their proximity to stuff we already owned," said Marcus, who played a lead role in structuring the transaction.[89] What's more, Adelphia's Los Angeles system would give Time Warner Cable a major presence in the second-largest media and advertising market in the country. The Dallas system and Midwest properties also would raise Time Warner Cable's profile in

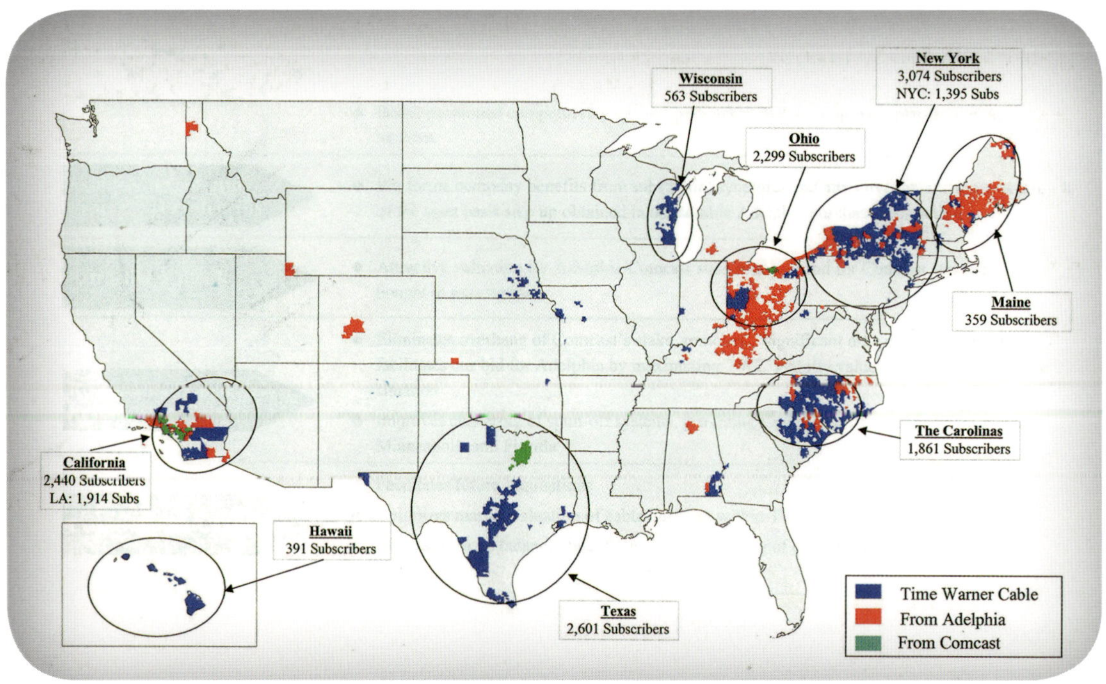

RIGHT
The acquisition of Adelphia properties and the related swaps with Comcast, which closed in July 2006, substantially boosted Time Warner Cable's presence in key regions, as illustrated on this map produced at the time to show the anticipated overlay of the properties with Time Warner Cable's existing service areas.

Tom Mathews, executive vice president of human resources, oversaw the HR implications of integrating the operations acquired in the Adelphia/Comcast deal and adding 11,000 new employees to Time Warner Cable's workforce.

Texas and Ohio. Time Warner Cable was in New York City, of course, but company executives and national advertisers were very much aware that rival Comcast had a presence in eight of the top 10 designated market areas.

Time Warner Cable and Comcast had worked together well in restructuring the TWE partnership following Comcast's purchase of the AT&T systems. Taking a similar approach to bidding for Adelphia would enable them to share the cost of such a major undertaking and obtain properties that best fit their existing footprints. It would also finish the job they started with the TWE restructuring in unwinding Comcast's ownership in Time Warner Cable.

"Ultimately," said Marcus, "what turned out to be the most elegant solution was a joint bid for Adelphia with Comcast that also enabled us to do a series of trades and swaps of their equity in us for assets. ... We were able to, in one fell swoop, significantly expand our footprint in a strategic and tax-efficient way, and extricate ourselves from the relationship with Comcast."[90]

In April 2005, Time Warner Cable and Comcast announced that they had reached definitive agreements to acquire substantially all of the assets of Adelphia. At the July 31, 2006, closing, Time Warner Cable picked up three million of Adelphia's roughly five million basic cable subscribers, and nearly a half-million from Comcast after accounting for all inter-company swaps, boosting its total subscriber base to about 14.4 million.[91] As a result of these acquisitions—and the restructuring that had begun prior to the Adelphia deal—85 percent of Time Warner Cable subscribers were grouped in five large, well-defined clusters in New York, Texas, California, Ohio, and the Carolinas.

Before and after closing, the complexities of integrating the new properties created herculean challenges throughout the company. "This was one of the most complex business transactions ever. It was complicated by the ever-shifting timing and whims of the bankruptcy court. On top of that, we had to create new business functions like SEC accounting and reporting, investor relations, and treasury as we prepped to become a separately traded public company, even though Time Warner would be our dominant shareholder," said Tom Mathews, executive vice president of human resources. "Not insignificantly, we on-boarded about 11,000 new employees who joined TWC after closing. I don't think there was a single function in the company that wasn't somehow pulled into the transition. And that was even before facing the challenges of integrating the millions of new customers we started serving."[92]

Time Warner Cable won its second Emmy Award in 2005 for its Start Over® digital video technology, following the 1994 Emmy for its work in fiber-optics, and in 2008 would win its third Emmy for its video on demand contributions to digital video technology.

Start Over® Emmy®

Time Warner Cable won its second Emmy® Award for the 2005 launch of Start Over® digital video technology. Start Over—developed as part of the Mystro project under Jim Chiddix and his group of engineers—made it possible for digital cable customers to restart a television program in progress without having to plan ahead of time or use any in-home recording devices. A software upgrade to the existing video on demand platform and installed digital set-top boxes enabled the new technology.

The new service was launched in October 2005 in the South Carolina division. The rollout began the following year in Rochester, New York; San Antonio, Texas; Greensboro, North Carolina; and Hawaii, and within a few years was available throughout the company.

"I Felt I Was Truly Valued by This Company"

Amanda Pruitt
Customer Care Center Lead
Kettering, Ohio

I have been with Time Warner Cable since August 2008, when I joined the Kettering call center as a call center representative. Before that, I had worked in a bank and in the fast-food industry.

After six weeks of training, when I was ready to hit the ground running, Mother Nature decided to test if I was truly ready. Hurricane Ike hit southwest Ohio hard, knocking out power, phone, cable, and, in some places, water through the entire region. People still needed to check voice mail from friends and family outside of the area who were calling them to make sure they were okay.

I spent my first week on the floor resetting voice mail PINs and helping customers check messages from cell phones. At one point, I thought to myself, "This isn't for me," but I came to work the next day, and our managers supplied us with food and lamps that read "I survived Ike."

I felt I was truly valued by this wonderful company and decided right then and there that I will retire from TWC. In less than a year, I got promoted to call center lead.

While I loved it, I missed working face to face with customers, so I took the position of care center lead in December 2010. Switching my job position has been the most challenging. Now from listening to our customers' pain, I can see it. But I'm loving it because I can do more to ease their pain faster and make them feel confident they made the right choice with TWC as their service provider.

My (customer care) reps are my "customers." I need to respect them, show empathy, and build rapport the same way they need to with our customers. Since moving to the care center, I've found TWC's values being lived every day. The reps have welcomed me to the team with open arms and are willing to show me the ropes and make sure I know everything to be successful. I had thought I was going to have to start from scratch and prove myself all over again, but I feel like I just got married into a wonderful new family.

New Stock Listing

In closing the deal, Time Warner Cable redeemed Comcast's 17.9 percent interest in Time Warner Cable and exchanged Comcast's 4.7 percent holding in TWE for various systems and cash. The two operators also swapped certain systems to create more efficient geographic clusters.

A subsidiary of Time Warner Cable paid Adelphia $8.9 billion in cash, and Time Warner Cable issued Time Warner Cable common stock to Adelphia representing approximately 16 percent of Time Warner Cable's equity, after adjusting for the related transactions with Comcast.[93] Under the terms of the deal, Time Warner Cable was required to list its stock on an exchange, ensuring that the Adelphia creditors would have a market in which to sell their stock.

When these transactions closed in 2006, Adelphia was still in bankruptcy. That meant that, instead of immediately distributing the stock to Adelphia's creditors, 16 percent of Time Warner Cable remained tied up in the bankruptcy proceeding. Finally, in February 2007, the bankruptcy process was completed, and the Time Warner Cable shares were distributed to Adelphia's creditors. In March, the cable company's common stock was once again trading on the New York Stock Exchange under the "TWC" symbol, with Time Warner owning 84 percent of the shares (representing a 90.6 percent voting interest) and 16 percent owned by the public.[94]

Maturing Industry

The pressure exerted by nationwide phone and satellite competitors to operate on a similar scale was keenly felt in Time Warner Cable's advertising and sales arenas during this period. The company used the process of restructuring into regions to position itself with advertisers as a nationwide presence. That was okay but it didn't go far enough, decided Joan Gillman, head of media sales. She felt Time Warner Cable and the other top cable operators needed to band together and present a common platform for advertisers.

Cable advertising, formerly a heavy growth industry, was clearly maturing by mid-decade. Not enough new content was launching, and programming that was launching wasn't getting a rating point from Nielsen, which was needed to attract advertisers. Phone companies, notably AT&T and Verizon with its FiOS product, and the satellite providers were competing for household video viewers. All of this was going on "at a time when the Internet was getting healthier and media was fragmenting and marketers were looking to move share to other media, away from traditional media, because traditional media was less accountable," Gillman said.[95]

Britt and Hobbs gave Gillman the green light to approach Comcast and Cox about jointly funding a McKinsey study of the opportunities and challenges of cable industry advertising. The study was completed by April 2007. Shortly thereafter, the six largest operators agreed to form a group, with Landel Hobbs and Steve Burke, Comcast chief operating officer, as co-chairs, Gillman the business lead, and Arthur Orduna of Bright House as the technical lead.[96] In the spring of 2008, the cable industry advanced advertising initiative was created as Canoe Ventures.

"The Rationale for Time Warner to Own Us Was Getting Thinner"

The fact that Time Warner Cable was a public company by the spring of 2007, albeit with 84 percent of the stock still owned by Time Warner, added momentum to an issue that had been discussed among Dick Parsons and Time Warner corporate executives and Britt and his cable team since the early years of the decade: spinning off the cable company as a standalone business.

As Britt recalled, "It had become clear to me early in the 2000 decade that the cable business was evolving to one that was more like a telecom business and more about technological platform infrastructure, and perhaps less of an entertainment business than it had been. The rationale for Time Warner to own us was getting thinner, I thought."[97]

Parsons recalled the spinoff issue coming up as early as 2002, shortly after he was named CEO.[98] Britt said to Parsons at that time that "I think in the long run, you should spin this company off."[99] Parsons had pressing problems, however. The first was a Securities and Exchange Commission investigation and a rash of lawsuits over AOL accounting practices that surfaced that year and dragged on for a number of years.

Joan Gillman, executive vice president of Time Warner Cable and president of media sales, set about reorganizing the company's sales efforts and the industry's joint approach to advertising.

Shortly after Time Warner Cable went public in 2007, the company's board of directors joined members of the senior management team at the New York Stock Exchange and marked the occasion by ringing the opening bell. Shown from left are board member Michael Lynne, Tom Robey, John Martin and Rob Marcus, board members Nick Nicholas and Wayne Pace, Landel Hobbs, John Thain of the NYSE, board member Don Logan, Glenn Britt, board members Thomas Castro, David Chang, Jim Copeland and Carole Black, Marc Lawrence-Apfelbaum, and board member Peter Haje.

Path to Independence

A change at the top of parent Time Warner helped turn the idea of spinning off the cable business a reality. Time Warner chief operating officer Jeff Bewkes, who led HBO for years and in 2002 assumed responsibility for all of Time Warner's studio and programming businesses, was named in November 2007 to succeed Dick Parsons as CEO of Time Warner. He clearly agreed with Britt that the corporate parent was primarily a content company. The stage was set for the spinoff to take place.

Separation Agreement

The separation, which was announced publicly on May 20, 2008, and finalized 10 months later in March 2009, was an important milestone for Time Warner Cable executives. But Britt and Marcus knew that there was still much work to be done. They and Time Warner Cable's 47,000 other employees were leaving the Time Warner nest and flying into the most competitive cable environment ever. And while the economy was clearly showing signs of stress as of spring 2008, they had no idea that they were about to hit the worst economic downdraft since the Great Depression.

Glenn Britt, center, unveiled the Home to the Future exhibit at the Time Warner Center in New York, showcasing the company's industry-leading technologies in 2007. Jeff Bewkes, left, CEO of Home Box Office, was named to succeed Dick Parsons, right, as CEO of Time Warner late that year, paving the way for the spinoff of Time Warner Cable, which was announced in early 2008.

"How Does It Get Better Than That?"

Kathy Oda
Senior Project Manager
Billing Technology Services Group
Charlotte, North Carolina

I'm a senior project manager in our technology services group. Cable has been a part of my professional life for the last 22-plus years. I joined Warner Cable in 1988 as a computer Operator in Dublin, Ohio, then moved to the Denver office in the same capacity. Later, I supported our divisions as part of the Corporate Support Desk. When I joined, QUBE was cutting-edge technology, then came Road Runner to transform our industry, and now we're our own Local Exchange Carrier. How many other industries have this kind of history, technology, and longevity?

We've gone from being a siloed organization to working together for the greater good of our customers. As our company has grown, so have the number of applications and products to make life easier for our customer service reps. We provide a more robust product line for our customers. Over the past three to four years, I've seen the formalization of the cross-functional team concept for different projects, and new product launches have helped everyone reach the same goal.

Prior to joining the company, I worked for a large bank, and I can see the difference. It makes me really appreciate how lucky we are to work for a company that has evolved over the years, but what's constant is that we're still family. I'm proud of the products we offer and how we continue to evolve, grow, and adapt to the challenges of our market. When I talk to employees of other cable companies, I realize that although we're a large company, at heart we're still an entrepreneurial one. New ideas are welcomed and encouraged, and excellence rewarded. How does it get better than that?

Cable Dividend

Among the key points to be negotiated in the months leading up to signing the final separation agreement was how the transactions would be structured and how the intertwined ownership would be unwound. Time Warner Cable's independent directors formed a special committee to consider the transaction, with James E. Copeland, Jr., as chair, along with Carole Black, Thomas Castro, David Chang, Peter Haje, and N. J. Nicholas, Jr., and its own financial and legal advisors. The committee played a key role in the negotiations, fully cognizant of its fiduciary role as the independent voice of the company and its shareholders who weren't associated with the parent.

Cable companies, with their business model based on a steady stream of subscription revenues, traditionally carry a significant amount of debt in order to finance system upgrades and the construction or purchase of new systems or businesses. Time Warner Cable wanted the ratio of its debt to its earnings before interest, taxes, depreciation, and amortization (EBITDA)—a widely used financial yardstick for the industry—to be optimal after financing the separation.[100]

True to the spirit of the relationship between the two companies, a mutually beneficial agreement was reached. Time Warner Cable agreed to pay a $10.9 billion dividend to Time Warner Cable shareholders, of which $9.25 billion went to Time Warner, as holder of 84 percent of the company's stock. Time Warner Cable planned to pay for the dividend out of an existing credit facility and a $9 billion two-year bridge loan from a syndicate of banks. With that additional borrowing, Time Warner Cable's total debt jumped to $24.2 billion,[101] temporarily pushing its debt to EBITDA ratio to 3.83.

After the dividend payment of $10.27 per share of Time Warner Cable stock, one of the steps in the separation-related transaction before Time Warner distributed the Time Warner Cable shares to its stockholders was a one-for-three reverse stock split. On March 16, 2009, the Time Warner Cable common stock was trading in the $25 per share range.[102]

Calm Before the Storm

In retrospect, it is hard to imagine a worse time than mid-2008 for a company to load up on billions in additional debt. The collapse of Bear Stearns in March spooked global credit markets, but market conditions moderated somewhat by the time the spinoff was announced. As it turned out, the summer of 2008 was the calm before the storm. "The issue was that we were taking on leverage, which was in excess of the leverage we thought we'd want to have on a steady state basis," said Marcus. "And we were doing it unbeknownst to us at a time that turned out to be least opportune to do it."[103]

In June 2008, the company took advantage of the summer lull and raised $5 billion by issuing medium- to long-term public debt to repay a portion of the two-year bridge loan.[104] And even after Lehman Brothers' bankruptcy in September 2008 riled world financial markets, Time Warner Cable was able to return to the credit markets in mid-November to borrow an additional $2 billion, despite the fact that its stock was trading near an all-time low.

"That was about the ugliest time that there was," to be tapping the public markets, Marcus said. "But it spoke to our relationships with the debt community," he added. "I credit our treasury team, who had developed a lot of credibility on the Street. We felt confident all the way through."[105]

Time Warner Cable senior management received kudos from investors for not trying to sugarcoat any of their announcements or forecasts during what quickly was recognized as the nation's most financially tumultuous period since the 1930s. That tell-it-like-it-is approach speaks volumes to investors and lenders, and generates an important level of trust.

"Once you control your own balance sheet and control your own relationship with your shareholders," said Marcus, "you have an ability to really live by the principles you most believe in, and that's how we operate today."[106]

The collapse of Lehman Brothers in September 2008 roiled the global financial markets and threw thousands out of work, yet Time Warner Cable's financial strength enabled it to borrow billions a few months later in conjunction with its separation from parent company Time Warner.

Chairman and CEO Glenn Britt, center, regularly speaks on behalf of Time Warner Cable and the cable industry before legislators and regulators. In November 2010, he addressed a Senate Commerce Committee hearing on the issue of retransmission consent, as did Joseph Uva, president and CEO of Univision Communications, left; and Thomas Rutledge, right, chief operating officer of Cablevision Systems.

Competitive Landscape

Even as it was weathering a dramatic, prolonged economic downturn, Time Warner Cable faced an equally unprecedented level of competition in its major geographic areas. Satellite competitors Dish Network and DirecTV remained well-funded, nationwide rivals. And while the satellite companies suffered initially when cable companies rolled out triple play video, Internet and phone service in the 2000s, they responded by forging deals with third-party providers to offer broadband and phone service, and focused on enhancing and heavily promoting their video products—including, in DirecTV's case, a successful, aggressive multi-year push on HD programming.

Verizon and AT&T also continue to challenge Time Warner Cable. The giant phone companies are pursuing different technological paths to bring broadband access and video services to the home. But they share the economies of scale, particularly in advertising, inherent in their "super-regional" footprints and remain outsized forces to be reckoned with.

Time Warner Cable suffered some quarterly net losses of video subscribers in recent quarters in 2010 and 2011, and the weak economy and slow-to-nonexistent housing growth is no doubt part of the reason. But some see another competitive threat looming in the form of so-called "over the top" providers of video programming via Internet connections. Netflix and YouTube are among the best known players in this field offering on-demand video. Such services still require a broadband connection, but alternatives to cable are available. While in early 2011 the number of customers "cutting the cord" for such services appears small, commentators view the potential for additional increased competition in the future as a real possibility.

While contending with over-the-top competition, Time Warner Cable also seized opportunities, where appropriate, to expand its geographic footprint. In August 2011, the company announced an agreement to buy closely held Midwest cable operator Insight Communications Company Inc. for $3 billion. Insight served more than 750,000 customers.

Rob Marcus Named President and Chief Operating Officer

In December 2010, Time Warner Cable announced that CFO Rob Marcus had been named president and chief operating officer, replacing COO Landel Hobbs. Glenn Britt cited Marcus' "intelligence, strategic insight

and clear understanding of consumers"[107] as strengths that would serve Time Warner Cable well as Marcus assumed his new responsibilities. Britt also thanked Hobbs for his service. "In particular, his leadership in operations and marketing has helped Time Warner Cable grow and evolve in the face of increased competition and technological change."[108]

Peter R. Haje, lead director of the Time Warner Cable Board and former general counsel at Time Warner, and someone who has worked with Marcus since he was a young outside attorney assigned to Time Warner,[109] said that "Rob's visionary and creative leadership have significantly influenced Time Warner Cable's standing as a publicly traded company and as a leader in the communications industry. The company is performing well competitively, operationally, and financially. I'm confident that, under the leadership of Glenn and Rob, it will continue to excel."[110]

Marcus remained acting CFO while a search was mounted to find a replacement. In May 2011, the company announced that Irene Esteves had been named Time Warner Cable's new CFO. She was formerly CFO at XL Group PLC, a global insurance and reinsurance company.

Commercial Opportunities

From a standing start nearly a decade ago, the commercial component of Time Warner Cable topped $1 billion in revenue in 2010 and is continuing to grow at a brisk pace in 2011. The company sees great potential for growth going forward, including selling its services to enterprises. Within this subset, a highly profitable business has mushroomed providing what's known as "back haul" service to cellular phone companies.

Time Warner Cable expanded its business services offering in April 2011 when it completed the acquisition of NaviSite, Inc. a provider of enterprise-class hosting, management application, messaging, and cloud services. With a potential market estimated by some at $25 billion, business services represents an opportunity that "could be bigger than the residential business, but that would mean it would have to be very, very successful," said Gerry Campbell.[111]

TOP LEFT
Rob Marcus, formerly chief financial officer, was named president and chief operating officer in December 2010 and works closely with chairman and CEO Glenn Britt to shape Time Warner Cable's future.

TOP
Some of Time Warner Cable's senior executives are pictured on the floor of the NCTA Cable Show 2011 in Chicago. From left to right: executive vice president of operations, East region Carol Hevey; chief communications officer Ellen East; chief strategy officer Peter Stern; chief financial officer Irene Esteves; chief technology officer Mike LaJoie; chairman and CEO Glenn Britt; executive vice president of operations, West region Bill Goetz; president and COO Rob Marcus; chief content and video officer Melinda Witmer; and executive vice president of architecture, development and engineering Mike Hayashi.

Ads and other marketing materials promoting Time Warner Cable's residential and business services over the past decade reflect the wide breadth of products developed over that period, as well as the ever-evolving approach to marketing products in a rapidly changing communications marketplace.

Make any room a TV room.
TWCable TV™ for iPad.®
Download it today — free.
Remote DVR Manager | Interactive Guide | Over 100 channels

NO LIMITS
FAST FORWARD
PAUSE
REWIND
PLAY
With Time Warner Digital Cable there are no limits...
TIME WARNER CABLE

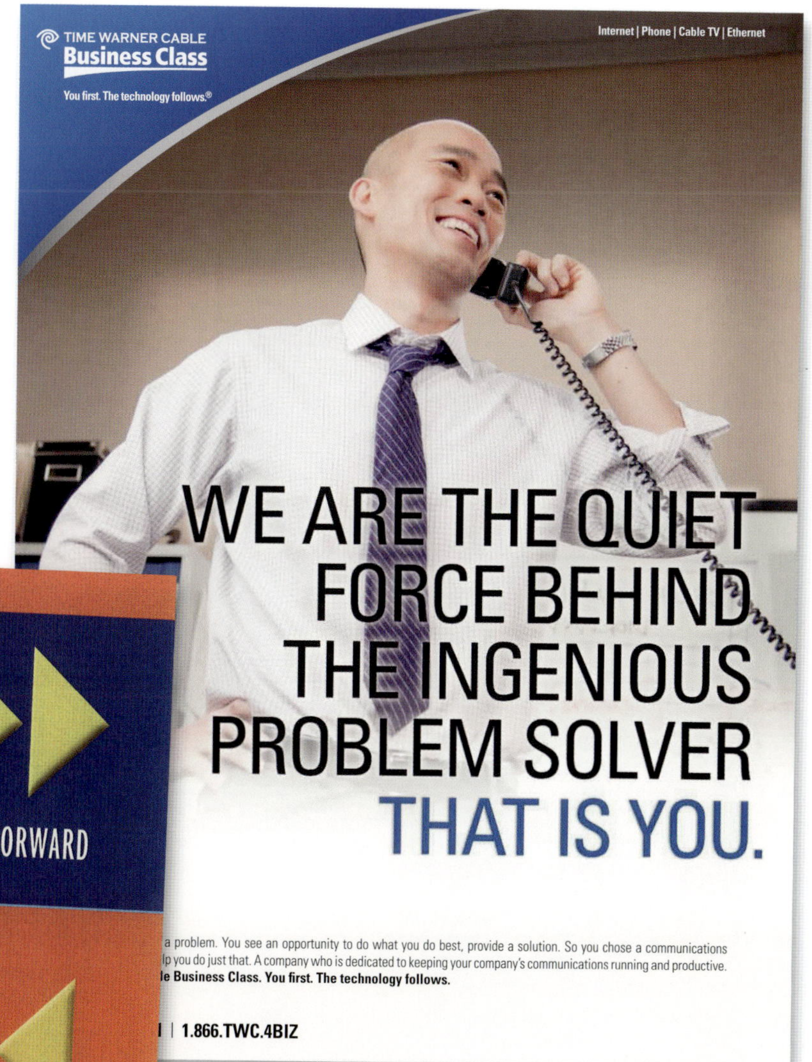
TIME WARNER CABLE
Business Class
You first. The technology follows.®
Internet | Phone | Cable TV | Ethernet
WE ARE THE QUIET FORCE BEHIND THE INGENIOUS PROBLEM SOLVER
THAT IS YOU.
1.866.TWC.4BIZ

Time Warner Cable's board of directors in 2011: front row, from left: Donna A. James, N.J. Nicholas, Jr., Glenn A. Britt, Wayne H. Pace, and Carole Black; middle row: Edward D. Shirley, Peter R. Haje, James E. Copeland, Jr., and Thomas H. Castro; back row: John E. Sununu, Don Logan, and David C. Chang.

"I See All of Our Ancestors"

Even as Time Warner Cable was grappling with external events such as the economic crisis and mounting competition, it was also engaged in a process of self-examination as it redefined who it was as a standalone business. What are the cultural building blocks of Time Warner Cable and the principles that drive the enterprise? The company had roughly four decades of experience under its tool belt informing how it perceived—and was perceived by—customers, employees, and its peers. "I think we have a culture of wanting to do things well, so there's a culture of excellence," said Britt. "That doesn't mean that we are always where we want to be. It doesn't mean everything is always perfect."[112]

One of the key strengths of the company, in fact, is that it draws from three distinct cultural traditions. "I see all of our ancestors," said Britt. "Clearly, there's a strong sense of what the original Time Inc. was about—values and culture. But there's a lot of the Warner culture, too," pushing the innovation and technology envelope, he added. "I'd like to think it's the best of both," Britt said. "There's also a lot of the original ATC culture, which was separate from Time. ... When you talk about how we present ourselves as a public company, that's probably the way Monty Rifkin would do it. I think we clearly have that set of values going back to ATC."[113]

Experience it. Use your smartphone to watch a video of Glenn Britt describing the ways in which Time Warner Cable's predecessors continue to influence the company's values today.

The Next Chapter

Glenn Britt and Rob Marcus lead Time Warner Cable's next chapter with the absolute certainty that major challenges and changes lie ahead. They are just as certain and enthusiastic about the opportunities the company is positioned to take advantage of. The company seeks to maintain its tradition of innovation leadership with the continued introduction of products that give customers greater control over the technology that Time Warner Cable delivers to their homes. It also continues its trend of leading the industry in understanding and serving well-defined groups of customers and their needs. And Time Warner Cable prides itself on the connections it forges with its workforce, and the connections they make with each other and customers in communities across America.

"I've always been proud of the people who make this company successful, and I continue to admire their ability to do that despite significant challenges," Britt said. Time Warner Cable's success ultimately comes down to "our ability to continue to attract and provide career opportunities for one of the most highly rated and motivated workforces in the cable industry," Britt added. "And to do that, the company will strive to live out its mission and values in a shared sense of purpose so that Time Warner Cable employees have more than a job. They have a frontline role in making communications industry history and writing the next chapter in the Time Warner Cable story."[114]

The dedication of and quality of service provided by Time Warner Cable's frontline employees continue to make them one of the company's most important competitive advantages.

Afterword

Every era in the history of Time Warner Cable has brought unique opportunities and challenges. When you're in those moments, every event can seem urgent and important. Yet what might seem noteworthy at one moment may end up only a blip when considered in the bigger context of history (if it even appears at all). That's why our recent past and the present day were probably the hardest to capture in this book.

As we write this today, in mid-2011, Time Warner Cable has been a standalone, independent company for more than two years. During this period of prolonged economic weakness, our company has returned significant value to shareholders in the form of a regular quarterly dividend and share repurchase. We're proud of that.

We're also proud of our ability, demonstrated throughout this book, to harness new and rapidly developing technologies to bring more relevance and value to our customers' lives. We are very focused on meeting their changing needs by giving them more control in ways that are simple and easy. At present, consumers' desire to view content on their own schedule and from any device they choose is influencing many of the product decisions and advancements we make as a company. Our recent launch of the TWCable TV app for the Apple iPad® is one example of how we're giving our customers more control over how, when and where they use our services and enjoy content.

Most important, this afterword gives us an opportunity to thank the 47,000 employees of Time Warner Cable, as well as the thousands more who have been part of our company and predecessor companies over the years. They keep our services fresh and current, reliable and relevant despite all the challenges of the day—whether that day was in 1968 or 2011.

Whatever decisions the company makes today and in the future, we are influenced not only by what's happening in the present, but by what has come before. As a company, we've learned a lot from many of the events chronicled in the previous pages and the decisions made by the company's leaders and their teams, even if they didn't realize at the time that they were creating a legacy that guides us today. We are grateful for the foundation they created and the lessons they continue to teach us.

Thank you,

Glenn Britt
Chairman and Chief Executive Officer

Rob Marcus
President and Chief Operating Officer

Acknowledgments

We would like to thank the many people who worked diligently and gave generously of their time and expertise to help create this history of Time Warner Cable. Chairman and CEO Glenn Britt and president and COO Rob Marcus, along with many other current and former leaders of the company listed here, graciously provided stories, memories and insights throughout development of the book. In particular, we would like to thank Dave O'Hayre, Time Warner Cable's former head of mergers and acquisitions strategy. Dave played an important role in helping shape the initial book concept and shepherd it through numerous iterations while also providing wise counsel on all things cable. If he didn't know the answer to a question about the company or the industry, he knew where it could be found. The staff of The Cable Center in Denver, including Lisa Backman, Brian Kenny, and Diane Rabson, was tremendously helpful. They provided numerous images and other content from their impressive archives, and they were gracious, accommodating, and patient throughout. We would also like to acknowledge and thank Ellen East, Time Warner Cable's chief communications officer, and Anthony Surratt, vice president of corporate communications, for their role in piloting the book project from start to completion.

Thanks to the following current and former Time Warner Cable officers and employees and industry leaders and advisors who were interviewed in the preparation of this book.

Satish Adige
Dick Aurelio
Tom Binning
Judy Braden
Glenn Britt
Julian Brodsky
Gerry Campbell
Jim Chiddix
Joe Collins
Jim Cottingham
Jimmy Doolittle
Ellen East
Aaron Fleischman
Steve Fry
Jack Gault
Alan Gerry
Joan Gillman
Bill Goetz
Jim Gray
Gus Hauser
Mike Hayashi
Carol Hevey
Landel Hobbs
Sam Howe
Veda James
Jeff King
Karen La Cava
Mike LaJoie
Brian Lamb
Lisa Lavelle
Marc Lawrence-Apfelbaum
Kevin Leddy
Jerry Levin
Rich Longwell
Jim Ludington
Gail MacKinnon
John Malone
Rob Marcus
Tom Mathews
Robert Miron
Dick Munro
Trygve Myhren
Nick Nicholas
Harriet Novet
Terry O'Connell
Kathy Oda
Dave O'Hayre
Stephen Pagano
Dick Parsons
Amanda Pruitt
Tom Reifenheiser
Monty Rifkin
Barry Rosenblum
Carl Rossetti
John Salinas
Paula Smith
Lenny Stern
Howard Szarfarc
Cesar Torres
June Travis
David Van Valkenburg
Louis Williamson
Lynn Yaeger

The following people played important roles in helping produce this publication.

Jeremy Asselin
Pat Berry
Jim Braun
David Christman
Dave Dasgupta
Christine Doney
Bonnie Hathaway
Michael Haught
Bill Jasso
Nanette Posman
Mike Quinn
Kate Rozen
Stacy Schmitt
Ann Shrewsbury
Steven Teplitz
Riina Tohvert
Carolyn Toner
Justin Venech
Susan Waxenberg
Gary Wengrofsky
Jeff Zimmerman

Notes

Chapter 1: Birth of an Industry

1 Monroe Rifkin, interview by Jim Keller, 13 May 1998, The Hauser Oral and Video History Collection, Barco Library/The Cable Center, Denver, transcript pp. 1-2.

2 Irving Kahn, interview by Marlowe Froke, Hauser Oral and Video History Collection, 10 July 1987 and 2 October 1988, Barco Library/The Cable Center, Denver, CO, transcript, p. 28, 90.

3 Rifkin, interview by Keller.

4 Parsons, Patrick R. *Blue Skies: A History of Cable Television*. Philadelphia, PA: Temple University Press, 2008, p. 23.

5 *The New York Times*, 8 April 1927, p. 1; Parsons, Patrick R. *Blue Skies: A History of Cable Television*, p. 30.

6 Parsons, Patrick R. *Blue Skies: A History of Cable Television*, p. 34.

7 Ibid., pp. 38-39, 41.

8 Ibid., p. 41.

9 Ibid., p. 38.

10 Ibid., p. 51.

11 *Newsweek*, 25 October 1948, p. 66.

12 "Young Monster," *Time*, 3 January 1949, p. 31.

13 Parsons, Patrick R. *Blue Skies: A History of Cable Television*, p. 47.

14 Ibid., pp. 48-49.

15 Ibid., p. 51.

16 For detailed discussion of CATV's roots, see Patrick R. Parsons' *Blue Skies: A History of Cable Television*, chapter 2.

17 Ibid., pp. 61-62.

18 Ibid., p. 64.

19 Ibid., p. 58.

20 Ibid., pp. 65-66.

21 Ibid., pp. 66-68.

22 Ibid., pp. 66-69.

23 Martin F. Malarkey, interview by Kathleen Hom, 22 August 1985, The Hauser Oral and Video History Collection, Barco Library/The Cable Center, Denver, CO.

24 Bill Daniels, interview by Max Paglin, 2 February 1986, The Hauser Oral and Video History Collection, Barco Library/The Cable Center, Denver, CO, transcript, p. 7.

25 Ibid., p. 5.

26 Ibid., p. 7.

27 Ibid. p. 8.

28 Parsons, Patrick R. *Blue Skies: A History of Cable Television*, pp. 102-103.

29 Ibid., p. 106.

30 Ibid., p. 107.

31 Daniels, interview by Paglin, transcript, p. 10.

32 Ibid., p. 19.

33 Ibid.

34 Rifkin, interview by Keller.

35 *Cablevision Plus*, 25 July 1983, p 41.

36 Rifkin, interview by Keller.

37 Ibid.

38 Hub Schlafly, interview by Rex Porter, 9 November 2000, The Hauser Oral and Video History Collection, Barco Library/The Cable Center, Denver, CO, transcript, p. 2.

39 Jack Gault, interview by Scott McMurray, 20 August 2010, The History Factory, Chantilly, VA, transcript, p. 2.

40 Ibid., p. 1.

41 Ibid., p. 16.

42 Ibid., p. 25.

43 Parsons, Patrick R. *Blue Skies: A History of Cable Television*, p. 178.

44 Ibid., pp. 119-120, 180.

45 Ralph J. Roberts, interview by Tom Southwick, July 2001, The Hauser Oral and Video History Collection, Barco Library/The Cable Center, transcript, p. 4.

46 Alan Gerry, interview by Tom Southwick, 6 June 2000, The Hauser Oral and Video History Collection, Barco Library/The Cable Center, Denver, CO, transcript. pp. 11, 13, 17, 18.

47 Parsons, Patrick R. *Blue Skies: A History of Cable Television*, p. 185.

48 Ibid., p. 186.

49 Ibid., p. 262.

50 Ibid., p. 263.

51 Ibid., p. 282.

52 Warner Bros. 1987 Annual Report; filing date 28 March 1988.

53 Rifkin, interview by Keller.

54 Ibid.

55 Ibid.

56 Monroe Rifkin, interview by Scott McMurray, 21 September, 2010, The History Factory, Chantilly, VA, transcript, p. 25.

57 ATC Corporate Minute Books, 7 June 1968-31 October 1968; SEC Form S1 filing, 9.24.68, p. 6.

58 ATC 1969 Annual Report, p. 1.

59 Parsons, Patrick R. *Blue Skies: A History of Cable Television*, p. 172.

60 Ibid., p. 173.

61 Rifkin, interview by Keller.

62 ATC 1969 Annual Report, p. 3.

Chapter 2: Coming of Age

1 Malarkey, interview by Hom, p. 2.

2 Parsons, Patrick R. *Blue Skies: A History of Cable Television*, pp. 279, 281-282.

3 Gerry Campbell, interview by Scott McMurray, 18 July 2010, The History Factory, Chantilly, VA, transcript, pp. 5-6.

4 Gus Hauser, interview by Scott McMurray, 4 October 2010, The History Factory, Chantilly, VA, transcript, pp. 4-5.

5 Gus Hauser, interview by Tom Southwick, 3 August 1999, The Hauser Oral and Video History Collection, Barco Library/The Cable Center, Denver, CO, transcript, p. 27,

6 Ibid., p. 5.

7 ATC company timeline, 1983, p.1.

8 David O'Hayre, interview by Scott McMurray, 13 July 2010, The History Factory, Chantilly, VA, transcript, p. 5.

9 John Gault, interview by Scott McMurray, 7 October 2010, The History Factory, Chantilly, VA, transcript, p. 1.

10 Barry Rosenblum, interview by Scott McMurray, 23 August 2010, The History Factory, Chantilly, VA, transcript, p. 4.

11 Jimmy Doolittle, interview by Scott McMurray, 10 August 2010, The History Factory, Chantilly, VA, transcript, p. 10.

12 ATC company timeline, p. 1.

13 "ATC Pursues CATV Leaders," *Rocky Mountain News*, 9 May 1971, p. 51.

14 David Kinley, interview by Liz Burke, 2001, Barco Library/The Cable Center, Denver, CO, transcript.

15 Ibid.

16 "A Profile of American Television and Communications Corporation," 1979, p. 1, The Cable Center, Denver, CO.

17 Joe Collins, interview by Bruce Weindruch, 30 September 2010, The History Factory, Chantilly, VA, transcript, pp. 10-11.

18 June Travis, interview by Scott McMurray, 10 November 2010, The History Factory, Chantilly, VA, transcript, p. 1.

19 Rifkin, interview by Keller, transcript, pp. 46-47.

20 Campbell, interview by McMurray, transcript, p. 11.

21 Travis, interview by McMurray, transcript, p. 2.

22 Ibid., p. 3.

23 ATC stock offering document, p. 14.

24 Ibid.

25 Rifkin, interview by McMurray, transcript, p. 26.

26 Collins, interview by Weindruch, transcript, p. 18.

27 Rifkin, interview by Keller, transcript, p. 26.

28 ATC company timeline, p. 1.

29 Rifkin, interview by McMurray, transcript, p. 27.

30 Collins, interview by Weindruch, transcript, p. 12.

31 Ibid., p. 7.

32 Parsons, Patrick R. *Blue Skies: A History of Cable Television*, p. 289.

33 Campbell, interview by McMurray, pp. 10-11.

34 Ibid., p. 7.

35 Parsons, Patrick R. *Blue Skies: A History of Cable Television*, pp. 243-244.

36 "Panel Would Lift Curbs on Cable TV," *New York Times*, 10 December 1968, p. 1.

37 Smith, Ralph Lee. *The Wired Nation: Cable TV: The Electronic Communications Highway.* New York: Harper Colophon Books, 1970.

38 Warner Cable Corp., presentation to N.Y. CATV Analysts Group, 30 September 1976. Barco Library/The Cable Center, Denver, CO.

39 Parsons, Patrick R. *Blue Skies: A History of Cable Television*, p. 265.

40 Jim Chiddix, interview by Rex Porter, 30 September 1999, The Hauser Oral and Video History Collection, Barco Library/The Cable Center, Denver, CO, transcript, pp. 8-9.

41 Parsons, Patrick R. *Blue Skies: A History of Cable Television*, pp. 276-277.

42 Ibid., p. 315.

43 Rifkin, interview by Keller, transcript, p. 28.

44 Doolittle, interview by McMurray, transcript, pp. 1-4.

45 Parsons, Patrick R. *Blue Skies: A History of Cable Television*, pp. 282-283.

46 Glenn Britt, interview by Bruce Weindruch, 28 July 2010, The History Factory, Chantilly, VA, transcript, p. 7.

47 ATC company timeline, p. 2.

48 Rifkin, interview by Keller, transcript, pp. 29-30.

49 Richard Munro, interview by Scott McMurray, 10 September 2010, The History Factory, Chantilly, VA, transcript, p. 5.

50 Rifkin, interview by Keller, transcript, p. 30.

51 Ibid.

52 Ibid., p. 31.

53 Rifkin, interview by McMurray, transcript, p. 29.

54 David Van Valkenburg, interview by Scott McMurray, 15 November 2010, The History Factory, Chantilly, VA, transcript, pp. 2-3.

55 Parsons, Patrick R. *Blue Skies: A History of Cable Television*, pp. 310-311.

56 Ibid., pp. 312-313.

57 Munro, interview by McMurray, transcript, p. 10.

58 Parsons, Patrick R. *Blue Skies: A History of Cable Television*, p. 313.

59 Ibid.

60 Britt, interview by Weindruch, 28 July 2010, transcript, p. 26.

61 Ibid., p. 28.

62 Nick Nicholas, interview by Bruce Weindruch, 2 November 2010, The History Factory, Chantilly, VA, transcript, p. 51.

63 Parsons, Patrick R. *Blue Skies: A History of Cable Television*, pp. 284-285.

64 Jerry Levin, interview by Bruce Weindruch, 9 November 2010, The History Factory, Chantilly, VA, transcript, p. 17.

65 Gerry, interview by Southwick, transcript, pp. 24, 47-48.

66 Parsons, Patrick R. *Blue Skies: A History of Cable Television*, pp. 298-299.

67 Nick Nicholas, interview by Bruce Weindruch, 2 November 2010, The History Factory, Chantilly, VA, transcript, pp. 47-48.

68 Parsons, Patrick R. *Blue Skies: A History of Cable Television*, p. 337.

69 Levin, interview by Weindruch, transcript, pp. 32-33.

70 25th Anniversary of Satellite Launch, interview by Lamb, Hauser Oral and Video History Collection, Barco Library/The Cable Center, Denver, CO, transcript, p. 10.

71 Parsons, Patrick R. *Blue Skies: A History of Cable Television*, p. 339.

72 Jeff King, interview by Scott McMurray, 12 August 2010, The History Factory, Chantilly, VA, transcript, pp. 11-12.

73 Ibid., p. 12.

74 Ibid., p. 13.

75 Nicholas, interview by Weindruch, transcript, p. 57.

76 Nicholas, interview by Nelson, transcript, p. 17.

77 Ibid., p. 17.

78 Ibid., p. 19.

79 "HBO: Point Man for an Industry Makes It into the Clear," Broadcasting, 17 October 1977, p. 51.

80 Robichaux, Mark. *Cable Cowboy: John Malone and the Rise of the Modern Cable Business.* Hoboken, NJ: John Wiley & Sons, 2002, p. 52.

81 Collins, interview by Weindruch, transcript, pp. 24-25.

82 Parsons, Patrick R. *Blue Skies: A History of Cable Television*, pp. 380-381.

83 Ibid., pp. 381-382.

84 R. E. Ted Turner, interview by Robert Maxwell, Hauser Oral and Video History Collection, Barco Library/Cable Center, Denver, CO, transcript, p. 2.

85 Parsons, Patrick R. *Blue Skies: A History of Cable Television*, pp. 89, 386.

86 Ibid., pp. 350-352.

87 Ibid., p. 452.

88 Gus Hauser, interview by Tom Southwick, Hauser Oral and Video History Collection, Barco Library/The Cable Center, Denver, CO, transcript, p. 51.

89 Parsons, Patrick R. *Blue Skies: A History of Cable Television*, pp. 452-454.

90 Brian Lamb, interview by Scott McMurray 13 October 2010, The History Factory, Chantilly, VA, transcript, p. 1.

91 Ibid., p. 4.

92 Ibid., pp. 4-5.

93 Campbell, interview by McMurray, transcript, p. 25.

94 Ibid., p. 26.

95 Hauser, interview by Southwick, transcript, p. 57.

96 Ibid., pp. 78-79.

97 Ibid., p. 81.

98 Nick Nicholas, interview by Steve Nelson, 11 June 2002, Hauser Oral and Video History Collection, Barco Library/The Cable Center, transcript, p. 20.

99 Rifkin, interview by Keller, transcript, p. 35.

100 Nicholas, interview by Nelson, transcript, p. 20.

101 Rifkin, interview by Keller, transcript, pp. 35-36.

102 Nicholas, interview by Nelson, transcript, p. 20.

103 Monty Rifkin, letter from Royal Little, 25 June 1987.

Chapter 3: Growth and Innovation

1 "Bring Plenty of Money," Allan Sloan, *Forbes*, 10 December 1979, pp. 49-54.

2 Tom Binning, interview by Scott McMurray, 16 September 2010, The History Factory, Chantilly, VA, transcript, pp. 14-15.

3 Jim Gray, interview by Scott McMurray, 26 October 2010. The History Factory, Chantilly, VA, transcript, p. 6.

4 Brown, Kathi Ann. *Wired to Win: Entrepreneurs of the American Cable Industry*. Fairfax, VA: Spectrum Publishing Group, 2003, p. 127.

5 Gray, interview by McMurray, transcript, p. 6.

6 Collins, interview by Weindruch, transcript, p. 28.

7 "Wire Wars: A Cable Fable," *Cablevision*, 15 December 1980, p. 75.

8 "The Franchise Story," Fred Dawson, *Cablevision*, 19 May 1980, p. 3.

9 "Wire Wars: A Cable Fable," pp. 75, 86.

10 PR Newswire, 9 December 1981; Associated Press, 11 December 1981.

11 Carl Rossetti, interview by Bruce Weindruch, 27 July 2010, The History Factory, Chantilly, VA, transcript, pp. 25-26.

12 Parsons, Patrick R. *Blue Skies: A History of Cable Television*, pp. 411-412.

13 Rossetti, interview Weindruch, transcript, pp. 26-27.

14 Ibid., p. 10.

15 Ibid., p. 4.

16 Ibid., p. 7.

17 Ibid., pp. 9-10; Chiddix, interview by Porter, transcript, pp. 15, 25, 41.

18 Britt, interview by Bruce Weindruch, 1 October 2010, The History Factory, Chantilly, VA, transcript, p. 3.

19 Britt, interview by Bruce Weindruch, 17 September 2010, The History Factory, Chantilly, VA, transcript, p. 6.

20 "The Franchise Story," p. 63.

21 "Franchise Flights," Hugh Panero, *Cablevision Plus*, 25 January 1982, p. 18.

22 Britt, interview by Weindruch, 28 July 2010, transcript, pp. 48-49.

23 Ibid.

24 Britt, interview by Weindruch, 17 September 2010, transcript, pp. 8-9.

25 "The Franchise Story," p. 61.

26 Parsons, Patrick R. *Blue Skies: A History of Cable Television*, p. 408; "Wire Wars: A Cable Fable, p. 75; Munro, interview by McMurray, transcript, pp. 18-19.

27 "Bring Plenty of Money," pp. 49-54.

28 Terry O'Connell, interview by Scott McMurray, 7 October 2010, The History Factory, Chantilly, VA, transcript, p. 1; *The New York Times*, 7 April 1982, p. 15.

29 O'Connell, intervew by McMurray, transcript, p. 2.

30 "Wire Wars: A Cable Fable," p. 75.

31 Gray, interview by McMurray, transcript, p. 6.

32 "Wire Wars: A Cable Fable," p. 75.

33 http://www.enquirer.com/century/ loc_cincinnatis_century11.html, accessed 5 November 2010.

34 Gray, interview by McMurray, transcript, pp. 5-7.

35 "The Politics of Playboy," Stephen Koff, *Cincinnati Magazine*, October 1983, p. 85.

36 "Franchise Flights," p. 17.

37 ATC company timeline, 1983, p. 4.

38 "Franchise Flights," p. 17.

39 ATC company timeline, 1983, p. 4.

40 "Franchise Flights," p. 17.

41 Ibid., p. 18.

42 Rifkin, interview by McMurray, transcript, p. 47.

43 Ibid., pp. 47-48; ATC company timeline, 1983, p. 4.

44 Hauser, interview by Southwick, transcript, pp. 26-27.

45 ATC company timeline, 1983, p. 4.

46 Ibid.

47 Gault, interview by McMurray, 7 October 2010, transcript, p. 2.

48 Myhren, interview by McMurray, transcript, p. 20.

49 Britt, interview by Weindruch, 17 September 2010, transcript, p. 27.

50 O'Hayre, interview by McMurray, transcript, p. 17.

51 ATC company timeline, 1983, p. 4.

52 Britt, interview by Weindruch, 17 September 2010, transcript, p. 27.

53 Ibid., p. 28.

54 Campbell, interview by McMurray, transcript, p. 29.

55 ATC company timeline, 1983, p. 4.

56 *InSync* newsletter, April 1982, p. 6.

57 Myhren, interview by McMurray, transcript, p. 20.

58 Parsons, Patrick R. *Blue Skies: A History of Cable Television*, p. 410.

59 Ibid., pp. 411, 413; John Malone, interview by Trygve Myhren, 22 October 2001, The Hauser Oral and Video History Collection, Barco Library/The Cable Center, Denver, CO.

60 Kevin Leddy, interview by Scott McMurray, 20 July 2010, The History Factory, Chantilly, VA transcript, p. 21.

61 Parsons, Patrick R. *Blue Skies: A History of Cable Television*, p. 411.

62 Myhren, interview by McMurray, transcript, p. 22.

63 Leddy, interview by McMurray, transcript, p. 21.

64 Parsons, Patrick R. *Blue Skies: A History of Cable Television*, pp. 477-478.

65 Hauser, interview by Southwick, transcript, p. 98; Malone, interview by Myhren.

66 "Heritage Project," David O'Hayre, p. 5.

67 Parsons, Patrick R. *Blue Skies: A History of Cable Television*, p. 517.

68 O'Hayre, interview by McMurray, transcript, pp. 17-18.

69 Parsons, Patrick R. *Blue Skies: A History of Cable Television*, p. 543.

70 Myhren, interview by McMurray, transcript, p. 22.

71 Ibid., pp. 22-23.

72 Parsons, Patrick R. *Blue Skies: A History of Cable Television*, pp. 513-514.

73 Ibid., pp. 5-6 ; O'Hayre, interview by McMurray, transcript, p. 6.

74 O'Hayre, interview by McMurray, transcript, p. 6.

75 Ibid.

76 Parsons, Patrick R. *Blue Skies: A History of Cable Television*, pp. 520-521.

77 O'Hayre, interview by McMurray, transcript, p. 6.

78 Parsons, Patrick R. *Blue Skies: A History of Cable Television*, p. 371.

79 Chiddix, interview by Porter, transcript, pp. 10-11.

80 Jim Chiddix, interview by Scott McMurray, 26 August 2010, The History Factory, Chantilly, VA, transcript, p. 46.

81 Ibid., pp. 2-3.

82 Ibid.

83 Ibid.

84 Ibid, pp. 23-24.

85 Britt, interview by Weindruch, 17 September 2010, transcript, p. 31.

86 Ibid., pp. 24-25.

87 Chiddix, interview by Porter, transcript, pp. 34-35.

88 Ibid., p. 18.

89 Ibid., p. 19.

90 Parsons, Patrick R. *Blue Skies: A History of Cable Television*, pp. 430-431.

91 Britt, interview by Weindruch, 17 September 2010, transcript, p. 29.

92 Collins, interview by Weindruch, transcript, pp. 31-32.

93 Britt, interview by Weindruch, 17 September 2010, transcript, p. 30.

94 Myhren, interview by McMurray, transcript, pp. 13-14.

95 Ibid., pp. 18, 24.

96 Bruck, Connie. "The World of Business Deal of the Year." *The New Yorker*, 8 January 1990, p. 66.

97 Parsons, Patrick R. *Blue Skies: A History of Cable Television*, p. 518; Munro, interview by McMurray, transcript, pp. 27-28; "The World of Business Deal of the Year," p. 68.

98 "The World of Business Deal of the Year," p. 66.

99 Britt, interview by Weindruch, 17 September 2010, transcript, p. 34.

100 Bruck, Connie. "The World of Business Strategic Alliances." *The New Yorker*, 6 July 1992, pp. 40-41.

101 Ibid., p. 72; Britt, interview by Weindruch, 17 September 2010, pp. 37-38.

102 "The World of Business Deal of the Year," pp. 66, 73.

103 Ibid., p. 72.

104 Ibid., pp. 73-74.

105 Munro, interview by McMurray, transcript, pp. 28-29.

106 Britt, interview by Weindruch, 17 September 2010, transcript, p. 39.

107 Ibid., p. 41.

Chapter 4: Taking the Lead

1 Lynn Yaeger, interview by Scott McMurray, 29 November, 2010, The History Factory, Chantilly, VA, transcript, p. 7.

2 Ibid.

3 Time Warner news release, 18 December 1991, p. 3.

4 Dick Aurelio, interview by Scott McMurray, 10 November 2010, The History Factory, Chantilly, VA transcript, pp. 11-12.

5 Ibid., p. 12.

6 Gault, interview by McMurray, transcript, pp. 6-7.

7 1990 ATC Annual Report p. 7.

8 Jim Chiddix, interview by Rex Porter, The Hauser Oral and Video History Collection, Barco Library/The Cable Center, Denver, CO, 30 September 1999, transcript, p. 12.

9 Ibid.

10 Ibid.

11 Time Warner news release, 7 March 1991, pp. 1-2.

12 Louis Williamson, interview by Scott McMurray, 8 November 2010, The History Factory, Chantilly, VA, transcript, p. 13.

13 Time Warner news release, 7 March 1991, pp. 1-2.

14 Time Warner news release, 18 December 1991.

15 Barry Rosenblum, interview by Scott McMurray, 23 August 2010, The History Factory, transcript, pp. 16-17.

16 Aurelio, interview by McMurray, transcript, p. 12.

17 Ibid., pp. 14-15.

18 *New York Post*, 25 April 1991, p. 3; *Multichannel News*, 13 May 1991, p. 55.

19 Gray, interview by McMurray, p. 13.

20 Ibid.

21 Ibid., pp. 13-14.

22 O'Connell, interview by McMurray, transcript, p. 8.

23 Landler, Mark and Dobrzynski, Judith H., et al. "Time Warner," *BusinessWeek*, 22 July 1991, pp. 70-71.

24 Bruck, Connie. "The World of Business Strategic Alliances," *The New Yorker*, 6 July 1992, pp. 47-48.

25 Ibid., p. 49.

26 Nicholas, interview by Weindruch, transcript, p. 118.

27 American Television and Communications Corp. Form 10-K, year ended December 31, 1991, Securities and Exchange Commission, p. 7.

28 Bruck, Connie. "The World of Business Strategic Alliances," p. 49.

29 Robert D. Marcus, interview by Bruce Weindruch, 1 October 2010, The History Factory, Chantilly, VA, transcript, pp. 9-10.

30 Bruck, Connie. "The World of Business Strategic Alliances," p. 51.

31 Ibid., p. 53.

32 "The Creator of Time Warner Steven J. Ross is Dead at 65," *The New York Times*, Obituaries, 21 December 1992.

33 Britt, interview by Weindruch, 1 October 2010, transcript, p. 25.

34 Ibid.

35 Time Warner 10-K form filed with the SEC, Washington, DC, for the year ending Dec. 31, 1993, p. I-1.

36 Skrzycki, Cindi. "Salvating at the Prospects of TCI-Bell Atlantic Deal," *Washington Post*, 9 November 1993; Garneau, George. "Cox, Southwestern Bell discuss major cable deal," *Editor & Publisher*, 13 November 1993.

37 Marcus, interview by Weindruch, transcript, p. 11.

38 Ibid.

39 Collins, interview by Weindruch, transcript, p. 53.

40 Leddy, Craig, "FSN: Vanguard of the Interactive Age," unpublished Time Warner Cable internal management report, 20 January 1999, p. 8.

41 Time Warner 10-K form filed with the SEC, Washington, DC, for the year ending Dec. 31, 1993, p. I-1.

42 Time Warner Entertainment Company L.P., Form 10-K for the year ending December 31, 1992, Securities and Exchange Commission, Washington, D.C., p. I-9.

43 Blodget, Henry. "Marc Andreessen: When We Founded Netscape, Everyone Thought We Were Nuts," *Business Insider*, 23 September 2009.

44 Roberts, Johnnie L., "Time Warner Plans a 'Superhighway of Electronics,'" *The Wall Street Journal*, 27 January 1993, p. 1.

45 Chiddix, interview by Porter, transcript, p. 13.

46 Chiddix, interview by McMurray, transcript, p. 27.

47 Ibid., pp. 27-28.

48 Britt, interview by Weindruch, 1 October 2010, transcript, p. 31.

49 Collins, interview by Weindruch, transcript, p. 46.

50 Britt, interview by Weindruch, transcript, pp. 31-32.

51 Roberts, Johnnie L., "Time Warner Plans a 'Superhighway of Electronics,'" p. 1.

52 Ibid.

53 Mike Hayashi, interview by Scott McMurray, 10 November 2010, The History Factory, Chantilly, VA, transcript, pp. 5-6.

54 Leddy, Craig, "FSN: Vanguard of the Interactive Age," unpublished Time Warner Cable internal management report, 20 January 1999, p. 9.

55 Ibid., p. 10.

56 Ibid., p. 21.

57 Collins, interview by Weindruch, transcript, p. 47.

58 Ibid; Leddy, Craig, "FSN: Vanguard of the Interactive Age," unpublished Time Warner Cable internal management report, 20 January 1999, p. 10.

59 Mhyren, interview by McMurray, transcript, p. 24.

60 1990 ATC Annual Report, p. 4.

61 Ibid.

62 Britt, interview by Weindruch, 1 October 2010, transcript, pp. 12-13.

63 Time Warner Form 10-K, filed with SEC, Washington, D.C, for year ending December 31, 1994, pp. I-25- I-26

64 Jimmy Doolittle, interview by Scott McMurray, 10 August 2010, The History Factory, Chantilly, VA transcript, p. 23.

65 Lenny Stern, interview by Scott McMurray, 3 November 2010, The History Factory, Chantilly, VA, transcript, p. 3.

66 Yaeger, interview by McMurray, transcript, p. 18.

67 Stern, interview by McMurray, transcript, p. 4.

68 Doolittle, interview by McMurray, transcript, p. 24.

69 Britt, interview by Weindruch, 23 November 2010, transcript, pp. 83-84.

70 Time Warner 2009 Annual Report, p. 8.

71 Ellen East, email to Scott McMurray, 3 August 2011.

Chapter 5: Transitions

1 Levin, interview by Weindruch, transcript, pp. 69–70.

2 Turner, Ted, with Bill Burke. *Call Me Ted*. New York: Grand Central Publishing, 2008, pp. 312–314.

3 Levin, interview by Weindruch, p. 70.

4 Turner, Ted, with Bill Burke. *Call Me Ted*, p. 319.

5 Saporito, Bill. "Time for Turner," *Time*, 21 October 1996; Time Warner 1998 10-K, p. F-39.

6 Turner, Ted, with Bill Burke, *Call Me Ted*, pp. 319–322.

7 Time Warner, Form 10-K for year ending December 31, 1998, Securities and Exchange Commission, Washington, D.C, p. F-39.

8 "Time Warner to Buy Turner Broadcasting," Corporate Growth Report, 11 September 1995, accessed online 11 January 2011, http://findarticles.com/p/articles/miqa3653/is199509/ain8732202.

9 Ted Turner, interview by Paul Maxwell, 28 November 2001, The Cable Center/Barco Library, The Hauser Video and Oral History Collection, transcript, p. 8.

10 Stetler, Brian. "Shaping a Network with Oprah's View," *The New York Times*, 31 December 2010, section B, p. 1.

11 Marc Lawrence-Apfelbaum, interview by Scott McMurray, 21 July 2010. The History Factory, Chantilly, VA, transcript, p. 18.

12 Turner, Ted, with Bill Burke. *Call Me Ted*, pp. 336–338; Kitman, Marvin. "Murdoch Triumphant," *Harper's Magazine*, November 2010, pp. 34–35.

13 Lawrence-Apfelbaum, interview by McMurray, 21 July 2010, transcript, p, 18.

14 Goodale, James C. "No Surprise in the City's Cable Loss," p. 2.

15 Marc Lawrence-Apfelbaum, interview by Scott McMurray, 6 January 2011, The History Factory, Chantilly, VA, transcript, p. 2.

16 Personal correspondence, Robert D. Joffe to Marc J. Apfelbaum, February 1998.

17 Robert Miron, interview with Scott McMurray, 11 October 2010, The History Factory, Chantilly, VA, transcript, p. 8.

18 Shapiro, Eben. "Big Names Beg Time Warner for Cable Channels," *The Wall Street Journal*, 21 April 1997, Sec. B, p. 1.

19 *The New York Times*, 22 April 1997, section B, p. 5,

20 Britt, interview by Weindruch, 1 October 2010, transcript, pp. 23-24; Time Warner 1999 10-K form, p. I-22.

21 Time Warner, Form 10-K for year ending December 31, 1996, Securities and Exchange Commission, Washington, D.C., p. I-2.

22 Miron, interview by McMurray, transcript, p. 9.

23 Ibid.

24 Alan Gerry interview with Scott McMurray, 15 October 2010, The History Factory, Chantilly, VA, transcript, p. 4–15

25 Ibid., p. 16.

26 Ibid., p. 17.

27 Ibid.

28 Time Warner, Form 10-K for year ending December 31, 1995, Securities and Exchange Commission, Washington, D.C, p. I-2.

29 Gail MacKinnon, interview by Scott McMurray, 4 January 2011, The History Factory, Chantilly, VA, transcript, p. 2.

30 Gail MacKinnon, interview by Scott McMurray, 18 August 2010, The History Factory, Chantilly, VA, p. 15.

31 Parsons, Patrick R. *Blue Skies: A History of Cable Television*, pp. 630–631.

32 Petition for Rulemaking before the Federal Communications Commission, 9 March 2010, pp. 14–15; MacKinnon, interview by McMurray, 4 January 2011, transcript, p. 3.

33 Britt, interview by Weindruch, 1 October 2010, transcript, p. 44.

34 Ibid., p. 45.

35 Britt, interview by Weindruch, 1 October 2010, transcript, p. 46.

36 Luftman, Mike, "Who knew in '93 that the two-way plant would sprout the HSD business?" *Cable World*, 10 March 2003.

37 Chiddix, interview by McMurray, transcript, p. 34.

38 Rossetti, interview by Weindruch, transcript, pp. 69–70.

39 Richard Parsons, interview by Scott McMurray, 24 August 2010, The History Factory, Chantilly, VA, transcript, p. 8.

40 Ibid.

41 Time Warner news release. "Road Runner Celebrates Its Tenth Anniversary," 19 September 2006; Rossetti, interview by Weindruch, transcript, p. 67.

42 Steve Fry, interview by Scott McMurray, 17 November 2010, The History Factory, Chantilly, VA, transcript, p. 11.

43 Ibid.

44 Ibid, p. 11; "Time Warner's Road Runner launches in Akron Canton Area," *PRNewswire*, 10 September 1996.

45 O'Connell, interview by McMurray, transcript, p. 8.

46 Rossetti, interview by Weindruch, transcript, pp. 67–68.

47 Britt, interview by Weindruch, 1 October 2010, transcript, p. 49.

48 Ibid., pp. 48–49.

49 Ibid., pp. 49–50.

50 Time Warner, Form 10-K for year ending December 31, 1998, p. F-36.

51 Kawamoto, Dawn and Jeff Pelline. "@Home IPO Seen as Hot Money, CNET News, 8 July 1997; http://news.cnet.com/Home-IPO-seen-as-hot-money/2100-1001_3-201258.html

52 Parsons, Patrick R. *Blue Skies: A History of Cable Television*, p. 655.

53 Lawrence-Apfelbaum, interview by McMurray, 21 July 2010, transcript, p. 15.

54 Time Warner, Form 10-K for year ending December 31, 1999, Securities and Exchange Commission, Washington, D.C, p. I-20; Road Runner Fortifies Industry Lead with Additions to Management Team," *Business Wire*, 21 June 1999.

55 Parsons, Patrick R. *Blue Skies: A History of Cable Television*, p. 655.

56 Robichaux, Mark. *Cable Cowboy*. Hoboken, NJ: John Wiley & Sons, 2002, p. 212.

57 Parsons, Patrick R. *Blue Skies: A History of Cable Television*, p. 637.

58 Britt, interview by Weindruch, 1 October 2010, transcript, p. 14.

59 cablelabs.com/anniversary/data.html, accessed 11 January 2011.

60 Collins, interview by Weindruch, transcript, pp. 71–72.

61 Robichaux, p. 192.

62 Parsons, Patrick R. *Blue Skies: A History of Cable Television*, p. 653.

63 Ibid.

64 Robichaux, p. 192.

65 Parsons, Patrick R. *Blue Skies: A History of Cable Television*, p. 653.

66 Hayashi, interview by McMurray, transcript, p. 8.

67 Time Warner, Form 10-K for year ending December 31, 1998, p. I-24.

68 Hayashi, interview by McMurray, transcript, p. 8.

69 Ibid., p. 9.

70 Mike Lajoie, interview with Scott McMurray, 27 October 2010, The History Factory, Chantilly, VA, transcript, pp. 7–8.

71 Ibid.

72 Ibid., p. 3.

73 "Greim, Lisa. "US West to Take Telecon Public; Initial Stock Offering Valued at $175 Million," *Rocky Mountain News*, 7 April 1998, p. 2B.

[74] Time Warner, Form 10-K for the year ending December 31, 1998, p. F-35.

[75] "IPO Report," 12 May, 1999, CBS Marketwatch.com.

[76] Time Warner, Form 10-K for the year ending December 31, 1999, p. I-21.

[77] Collins, interview by Weindruch, transcript, p. 62.

[78] Time Warner, Form 10-K for the year ending December 31, 1998, p. F-35.

[79] Time Warner, Form 10-K for the year ending December 31, 1999, p. I-19.

[80] Carol Hevey, interview by Scott McMurray, 25 August 2010, The History Factory, Chantilly, VA, transcript, pp. 10-11.

[81] Ibid., p. 11.

[82] Ibid.

[83] Collins, interview by Weindruch, transcript, p. 55.

[84] Robichaux, pp. 235-236.

[85] Ibid., pp. 238-239; Pelline, Jeff. "AT&T to Buy TCI for $48 Billion, 24 June 1998, *CNET News*.

[86] Jerry Levin, interview by Bruce Weindruch, 3 November 2010, The History Factory, Chantilly, VA, transcript, pp. 72-73.

[87] Ibid., pp. 73-74.

Chapter 6: Navigating Change

[1] LaJoie, interview by McMurray, transcript, p. 11.

[2] Time Warner, Form 10-K for year ending December 31, 1999, Securities and Exchange Commission, Washington, D.C, p. F-16.

[3] Ibid., p. I-20.

[4] Ibid., p. I-21.

[5] LaJoie, interview by McMurray, transcript, p. 11.

[6] Levin, interview by Weindruch, transcript, p. 76.

[7] Hu, Jim, "Case accepts blame for AOL-Time Warner debacle," 13 January 2005, zdnet.com; accessed online 12 January 2011.

[8] Lohr, Steve. "Media Megadeal: The Strategy; Medium for Main Street," *The New York Times*, 11 January 2000.

[9] Ibid., p. 31.

[10] Ibid.

[11] Chiddix, interview by McMurray, transcript, pp. 46-47.

[12] Peter Stern, interview by Scott McMurray, 20 July 2010, The History Factory, Chantilly, VA, transcript, pp. 1-6.

[13] Ibid., p. 8.

[14] Ibid.

[15] Turner, Ted, with Bill Burke. *Call Me Ted*, pp. 374-375.

[16] Ibid.

[17] Stern, interview by McMurray, transcript, p. 8.

[18] Klein, Alec. *Stealing Time*, New York: Simon & Schuster, 2003, pp. 186-187.

[19] Glenn Britt, interview by Bruce Weindruch, 23 November 2010, The History Factory, Chantilly, VA, transcript, pp. 16-17.

[20] Grice, Corey. "Time Warner-ABC dispute concerns cable industry, FCC," CNET News, 10 May 2000.

[21] Stern, interview by McMurray, transcript, p. 9; Klein, Alec. *Stealing Time*, p. 284.

[22] Stern, interview by McMurray, transcript, pp. 11-12.

[23] Chiddix, interview by McMurray, transcript, pp. 47-49.

[24] Parsons, interview by McMurray, transcript, p. 8.

[25] Ibid.

[26] Farrell, Mike and Kent Gibbons, "Levin opts to retire: Timing of AOL CEO's move a surprise," *Multichannel News*, 10 December 2001, p. 1.

[27] Britt, interview by Weindruch, 23 November 2010, transcript, p. 21.

[28] Parsons, Patrick R. *Blue Skies: A History of Cable Television*, Appendix A, p. 702.

[29] Britt, interview by Weindruch, 23 November 2010, transcript, p. 21.

[30] Parsons, Patrick R. *Blue Skies: A History of Cable Television*, p. 675.

[31] Britt, interview by Weindruch, 23 November 2010, transcript, p. 22; "Comcast Hangs In to Win Auction for AT&T Cable Unit," *Mergers and Acquisitions Journal*, 1 February 2002.

[32] Turner, Ted, with Bill Burke. *Call Me Ted*, p. 390.

[33] Ibid., pp. 374-375.

[34] Klein, Alec. *Stealing Time*, p. 270.

[35] Ibid., p. 271.

[36] Schiesel, Seth, and Geraldine Fabrikant, "Changing the Guard: The Overview; AOL Time Warner Gets a New Chief as Levin Departs," *The New York Times*, 6 December 2001, Section A, p. 1.

[37] Klein, Alec. *Stealing Time*, pp. 280-281.

[38] Ibid., p. 284.

[39] Ibid., p. 2.

[40] Ibid., p. 299.

[41] Miron, interview by McMurray, transcript, pp. 10-11.

[42] Ibid., p. 12.

[43] Ibid., p. 11; Time Warner Cable, Form 10-K for year ending December 31, 2005, Securities and Exchange Commission, Washington, D.C., pp. 29-31.

[44] Marcus, interview by Weindruch, 1 October 2010, transcript, p. 37.

[45] AOL Time Warner, Comcast press release dated 31 March 2003.

[46] Time Warner, Form 10-K for year ending December 31, 2005, Securities and Exchange Commission, Washington, D.C., pp. 29-31.

[47] Britt, interview by Weindruch, 23 November 2010, transcript, p. 28.

[48] "Time Warner Cable Announces New Organizational Structure," *Business Wire*, 12 October 2001; Britt, interview by Weindruch, 23 November 2010, transcript, pp. 32-33.

[49] Britt, interview by Weindruch, 23 November 2010, transcript pp. 32-33.

[50] Ibid., p. 29.

[51] Nanette Posman, email message to Scott McMurray, 1 February 2011.

[52] Harriet Novet group interview by Scott McMurray, 31 January 2011, The History Factory, Chantilly, VA, transcript, p. 5.

[53] Ibid., pp. 5-6.

[54] "New Heights of Excellence," *The CableCaster*, Winter 2001, p. 6.

[55] Ibid., p. 1.

[56] Ibid., p. 1.

[57] Ibid., p. 6.

[58] Ibid., p. 7.

[59] Novet group, interview by McMurray, transcript, p. 8.

[60] Ibid.

[61] Gerry Campbell, interview by Scott McMurray, 17 July 2010, The History Factory, Chantilly, VA, transcript, pp. 37-38.

[62] Ibid.

[63] Ibid., pp. 38-39.

[64] Ibid., p. 41.

[65] Ibid.

[66] Ibid., pp. 40-41.

[67] Ibid., p. 42.

[68] Ibid., p. 45.

[69] Baumgartner, Jeff, "Campbell looks Inside and Out to Dial up Time Warner's Aggressive VoIP Plans," *CedMagazine.com*, 1 July 2004.

[70] Campbell, interview by McMurray, transcript, p. 47.

[71] Ibid.

[72] Ibid., p. 48.

[73] Stern, interview by McMurray, transcript, pp. 17-18.

[74] Campbell, interview by McMurray, transcript, p. 50.

[75] Time Warner news release, 22 January 2004.

[76] Ibid., p. 77.

[77] Time Warner Cable, Form 10-K for year ending December 31, 2006, Securities and Exchange Commission, Washington, D.C., p. 10.

[78] Time Warner Cable, Form 10-K for year ending December 31, 2007, Securities and Exchange Commission, Washington, D.C., p. 7.

79 Stern, interview by McMurray, transcript, p. 21.

80 Britt, interview by Weindruch, 23 November 2010, transcript, pp. 30–31.

81 "3 Executives to Retire, Time Warner Cable Says, *Bloomberg News*, 15 January 2005.

82 Marcus, interview by Weindruch, 1 October 2010, transcript, p. 47.

83 Howe, interview by McMurray, transcript, p. 7.

84 Glenn Britt, email to Time Warner Cable employees, December 2007.

85 Bill Goetz and Carol Hevey, interview by Scott McMurray, 10 May 2011, The History Factory, Chantilly, VA, transcript, p. 9.

86 Ibid., p. 6.

87 Ibid.

88 Leonard, Devin, with Ann Harrington and Doris Burke. "The Adelphia Story," *Fortune*, 12 August 2002, pp. 136–148.

89 Marcus, interview by Weindruch, 1 October 2010, transcript, pp. 38–39.

90 Ibid.

91 Time Warner Cable, Form 10-K for year ending December 31, 2007, Securities and Exchange Commission, Washington, D.C., pp. 1–3.

92 Tom Mathews, email to Scott McMurray, 3 August 2011.

93 Time Warner press release, "Time Warner Cable and Comcast to Acquire Assets of Adelphia Communications," 21 April 2005.

94 Marcus, interview by Weindruch, 1 October 2010, transcript, p. 40.

95 Gillman, interview by McMurray, transcript, p. 8.

96 Ibid., pp. 9–10.

97 Britt, interview by Weindruch, 23 November 2010, transcript, pp. 36–37.

98 Parsons, interview by McMurray, transcript, p. 7.

99 Britt, interview by Weindruch, 23 November 2010, transcript, p. 37.

100 Satish Adige, interview by Scott McMurray, 30 November 2010, transcript, p. 10.

101 "Time Warner Cable announces $2B debt offering," 14 November 2008, Associated Press.

102 Adige, interview by McMurray, transcript, pp. 10–11.

103 Marcus, interview by Weindruch, 1 October 2010, transcript, p. 50.

104 "Time Warner Cable Prices a $5 Billion Debt Offering," 17 June 2008, Time Warner news release.

105 Marcus, interview by Weindruch, 1 October 2010, transcript, p. 50.

106 Ibid., p. 37.

107 Ibid.

108 Ibid.

109 Ibid., p. 2.

110 Time Warner Cable, company news release, "Time Warner Cable Announces Senior Management Changes, 14 December 2010.

111 Campbell, interview by McMurray, transcript, p. 1.

112 Britt, interview by Weindruch, 23 November 2010, transcript, p. 98.

113 Ibid., pp. 97–98.

114 Ibid.

Photo Credits

Introduction

iv-v Time Warner Cable
vii Time Warner Cable
viii-ix Time Warner Cable

Chapter 1: Birth of an Industry

Opening Photo Barco Library, The Cable Center; **2** Time & Life Pictures/Getty Images; **3** Barco Library, The Cable Center; **4** Barco Library, The Cable Center; **5** (top) Barco Library, The Cable Center; **5** (bottom) Barco Library, The Cable Center; **6-7** Barco Library, The Cable Center; **8** (left) Barco Library, The Cable Center; **8** (right) © Corbis; **9** New York World-Telegram and Sun Newspaper Photograph Collection (Library of Congress); **10** (top) Photo by NBC Television/Courtesy of Getty Images; **10** (bottom) Photo by Bob Peterson/Time Life Pictures/Getty Images; **11** (left) Barco Library, The Cable Center; **11** (right) Barco Library, The Cable Center; **12** © Corbis; 13 Barco Library, The Cable Center; **14** (left) Barco Library, The Cable Center; **14** (right) Barco Library, The Cable Center; **15** Time Warner Cable; **16** Barco Library, The Cable Center; **19** Barco Library, The Cable Center; **20** Barco Library, The Cable Center; **21** Barco Library, The Cable Center; **22** (left) Barco Library, The Cable Center; **22** (right) Barco Library, The Cable Center; **23** (left) Time Warner Cable/Warner Communications Inc.; **23** (right); Time Warner Cable; **24** Time Warner Cable; **25** Time & Life Pictures/Getty Images; **26** Time Warner Cable; **27** (top) Barco Library, The Cable Center; **27** (bottom) Barco Library, The Cable Center.

Chapter 2: Coming of Age

Opening Photo Barco Library, The Cable Center; **32** Barco Library, The Cable Center; **34** (left) Time Warner Cable; **34** (right) Barco Library, The Cable Center; **35** Barco Library, The Cable Center; **36** Getty Images; **37** (left) Time Warner Cable; **37** (right) Barco Library, The Cable Center; **38** (left) Time Warner Cable; **38** (right) Barco Library, The Cable Center; **39** Barco Library, The Cable Center; **40** (top) Getty Images; **40** (bottom) Time Warner Cable; **41** Barco Library, The Cable Center; **42** (left) Time Warner Cable; **42** (right) Getty Images; **43** Barco Library, The Cable Center; **44** Barco Library, The Cable Center; **45** Barco Library, The Cable Center; **46** (top) Time Warner Cable; **46** (bottom) Time Warner Cable; **47** Time Warner Cable/Time Inc.; **48** Time & Life Pictures/Getty Images; **49** Time Warner Cable/Time Inc.; **50** Barco Library, The Cable Center; **51** Barco Library, The Cable Center; **52** Barco Library, The Cable Center; **53** (top) Barco Library, The Cable Center; **53** (bottom) Barco Library, The Cable Center; **54-55** (all) Associated Press; **56** Time Warner Cable; **57** Getty Images; **58** (left) Barco Library, The Cable Center; **58** (right) Barco Library, The Cable Center; **59** Barco Library, The Cable Center; **60** (top) Barco Library, The Cable Center; **60** (bottom) Barco Library, The Cable Center; **61** (left) Barco Library, The Cable Center; **61** (right) Barco Library, The Cable Center; **62** (top left) qube-tv.com; **62** (top right) Barco Library, The Cable Center; **62** (bottom) Barco Library, The Cable Center; **63** Barco Library, The Cable Center; **64** Barco Library, The Cable Center; **65** (top) Barco Library, The Cable Center; **65** (bottom) Barco Library, The Cable Center.

Chapter 3: Growth and Innovation

Opening Photo Time Warner Cable/Time Inc.; **70** Barco Library, The Cable Center; **71** Barco Library, The Cable Center; **72** (left) Time Warner Cable; **72** (right) Time Warner Cable; **73** Time & Life Pictures/Getty Images; **74** (left) Time Warner Cable; **74** (right) Time Warner Cable; 75 Barco Library, The Cable Center; **76-77** Courtesy of AT&T Archives and History Center; **78** (top) Time Warner Cable; **78** (bottom) Time Warner Cable; **79** (left) Barco Library, The Cable Center; **79** (right) Time Warner Cable; **80** Barco Library, The Cable Center; **81** (top) Barco Library, The Cable Center; **81** (bottom) Barco Library, The Cable Center; **82** (left) SSPL via Getty Images; **82** (right) Barco Library, The Cable Center; **83** Barco Library, The Cable Center; **84** Time Warner Cable; **85** Barco Library, The Cable Center; **86** Barco Library, The Cable Center; **87** Barco Library, The Cable Center; **88** (top) Barco Library, The Cable Center; **88** (bottom) Time Warner Cable; **89** Barco Library, The Cable Center; **91** Barco Library, The Cable Center; **93** Time & Life Pictures/Getty Images; **94** Barco Library, The Cable Center; **95** Time Warner Cable; **97** (left) From *The New York Times*, March 5, 1989 © 1989 The New York Times. All rights reserved. Used by permission and protected by the Copyright Laws of the United States. The printing, copying, distribution, or retransmission of this Content without express written permission is prohibited; **97** (right) Time Warner Cable/Time Warner Inc.

Chapter 4: Taking the Lead

Opening Photo Barco Library, The Cable Center; **100** Barco Library, The Cable Center; **102** Barco Library, The Cable Center; **103** Barco Library, The Cable Center; **104** (left) Associated Press; **104** (right) Getty Images; **105** Barco Library, The Cable Center; **106** Time Warner Cable; **108** (top) Time Warner Cable; **108** (bottom) Time Warner Cable; **109** Associated Press; **110** © Corbis; **111** © Corbis; **112** © Corbis; **113** Associated Press; 114 © Corbis; 115 Barco Library, The Cable Center; **116** Barco Library, The Cable Center; **117** Time Warner Cable; **118** Barco Library, The Cable Center, **121** © Corbis, **122** Time Warner Cable; **123** (top left) Time Warner Cable; **123** (top right) Time Warner Cable; **123** (bottom) Time Warner Cable.

Chapter 5: Transitions

Opening Photo Time & Life Pictures/Getty Images; **126** Associated Press; **128** (left) © Corbis; **128** (right) Associated Press; **129** Time Warner Cable/Time Warner Inc.; **130** Associated Press; **131** Associated Press; **132** (left) Associated Press;

132 (top right) Time Warner Cable; **132** (bottom) Associated Press; **133** Time Warner Cable; **135** (left) Barco Library, The Cable Center; **135** (right) Time Warner Cable; **136** (left) Barco Library, The Cable Center; **136** (right) Associated Press; **137** Associated Press; **139** (top) Time Warner Cable; **139** (bottom) Time Warner Cable; **140** (top) Time Warner Cable; **140** (bottom) Time Warner Cable; **142** Time Warner Cable; **144** Associated Press; **145** Time Warner Cable; **146** Time Warner Cable; **147** Pivot Design; **148** Time Warner Cable; **149** Barco Library, The Cable Center; **150** Associated Press; **151** Time Warner Cable/Time Warner Inc.; **152** Associated Press.

Chapter 6: Navigating Change

Opening Photo Time Warner Cable; **156** Associated Press; **157** Associated Press; **159** Associated Press; **160** Associated Press; **161** NY Daily News via Getty Images; **162** (left) Time Warner Cable; **162** (right) © Corbis; **163** (top) Bloomberg via Getty Images; **163** (bottom) Time Warner Cable; **164** Associated Press; **165** Associated Press; **166** Associated Press; **167** Time Warner Cable; **169** Time Warner Cable; **170** Time Warner Cable; **171** Time Warner Cable; **172** Associated Press; **174** Time Warner Cable; **175** (top) Time Warner Cable; **175** (bottom) Time Warner Cable; **176** (top) Time Warner Cable; **176** (bottom) Time Warner Cable; **177** Time Warner Cable; **178** (top) Associated Press; **178** (bottom) Time Warner Cable; **179** (left) Time Warner Cable; **179** (right) Time Warner Cable; **180** Time Warner Cable; **181** Time Warner Cable; **182** Time Warner Cable; **183** (left) Time Warner Cable; **183** (right) Time Warner Cable; **184** Time Warner Cable; **185** From Time Magazine, October 13, 2008, © 2008 Time Inc. Used under license.; **186** Bloomberg via Getty Images; **187** (top) Time Warner Cable; **187** (bottom) Time Warner Cable; **188** (all) Time Warner Cable; **189** (all) Time Warner Cable; **190** Time Warner Cable; **191** Time Warner Cable.

Index

Spell Bound
CONTESTANT 1
CONTESTANT 2
CNN
CBS
Manhattan Cable TV
CABLEVision
Time Warner Cable
@TWCableHelp